DESIGN OF VERY HIGH-FREQUENCY MULTIRATE
SWITCHED-CAPACITOR CIRCUITS

THE KLUWER INTERNATIONAL SERIES IN ENGINEERING AND COMPUTER SCIENCE

ANALOG CIRCUITS AND SIGNAL PROCESSING
Consulting Editor: Mohammed Ismail. *Ohio State University*

Related Titles:

DESIGN OF WIRELESS AUTONOMOUS DATALOGGER IC'S
Claes and Sansen
Vol. 854, ISBN: 1-4020-3208-0
MATCHING PROPERTIES OF DEEP SUB-MICRON MOS TRANSISTORS
Croon, Sansen, Maes
Vol. 851, ISBN: 0-387-24314-3
LNA-ESD CO-DESIGN FOR FULLY INTEGRATED CMOS WIRELESS RECEIVERS
Leroux and Steyaert
Vol. 843, ISBN: 1-4020-3190-4
SYSTEMATIC MODELING AND ANALYSIS OF TELECOM FRONTENDS AND THEIR
BUILDING BLOCKS
Vanassche, Gielen, Sansen
Vol. 842, ISBN: 1-4020-3173-4
LOW-POWER DEEP SUB-MICRON CMOS LOGIC SUB-THRESHOLD CURRENT
REDUCTION
van der Meer, van Staveren, van Roermund
Vol. 841, ISBN: 1-4020-2848-2
WIDEBAND LOW NOISE AMPLIFIERS EXPLOITING THERMAL NOISE
CANCELLATION
Bruccoleri, Klumperink, Nauta
Vol. 840, ISBN: 1-4020-3187-4
SYSTEMATIC DESIGN OF SIGMA-DELTA ANALOG-TO-DIGITAL CONVERTERS
Bajdechi and Huijsing
Vol. 768, ISBN: 1-4020-7945-1
OPERATIONAL AMPLIFIER SPEED AND ACCURACY IMPROVEMENT
Ivanov and Filanovsky
Vol. 763, ISBN: 1-4020-7772-6
STATIC AND DYNAMIC PERFORMANCE LIMITATIONS FOR HIGH SPEED
D/A CONVERTERS
van den Bosch, Steyaert and Sansen
Vol. 761, ISBN: 1-4020-7761-0
DESIGN AND ANALYSIS OF HIGH EFFICIENCY LINE DRIVERS FOR Xdsl
Piessens and Steyaert
Vol. 759, ISBN: 1-4020-7727-0
LOW POWER ANALOG CMOS FOR CARDIAC PACEMAKERS
Silveira and Flandre
Vol. 758, ISBN: 1-4020-7719-X
MIXED-SIGNAL LAYOUT GENERATION CONCEPTS
Lin, van Roermund, Leenaerts
Vol. 751, ISBN: 1-4020-7598-7
HIGH-FREQUENCY OSCILLATOR DESIGN FOR INTEGRATED TRANSCEIVERS
Van der Tang, Kasperkovitz and van Roermund
Vol. 748, ISBN: 1-4020-7564-2
CMOS INTEGRATION OF ANALOG CIRCUITS FOR HIGH DATA RATE TRANSMITTERS
DeRanter and Steyaert
Vol. 747, ISBN: 1-4020-7545-6
SYSTEMATIC DESIGN OF ANALOG IP BLOCKS
Vandenbussche and Gielen
Vol. 738, ISBN: 1-4020-7471-9
SYSTEMATIC DESIGN OF ANALOG IP BLOCKS
Cheung and Luong
Vol. 737, ISBN: 1-4020-7466-2
LOW-VOLTAGE CMOS LOG COMPANDING ANALOG DESIGN
Serra-Graells, Rueda and Huertas
Vol. 733, ISBN: 1-4020-7445-X
CIRCUIT DESIGN FOR WIRELESS COMMUNICATIONS
Pun, Franca and Leme
Vol. 728, ISBN: 1-4020-7415-8
DESIGN OF LOW-PHASE CMOS FRACTIONAL-N SYNTHESIZERS
DeMuer and Steyaert
Vol. 724, ISBN: 1-4020-7387-9
MODULAR LOW-POWER, HIGH SPEED CMOS ANALOG-TO-DIGITAL CONVERTER
FOR EMBEDDED SYSTEMS
Lin, Kemna and Hosticka
Vol. 722, ISBN: 1-4020-7380-1

DESIGN OF VERY HIGH-FREQUENCY MULTIRATE SWITCHED-CAPACITOR CIRCUITS

Extending the Boundaries of CMOS Analog Front-End Filtering

by

Seng-Pan U

*University of Macau and Chipidea Microelectronics (Macau), Ltd.,
China*

Rui Paulo Martins

*University of Macau, China
and Technical University of Lisbon, Portugal*

and

José Epifânio da Franca

*Chipidea Microelectronics, S.A.
and Technical University of Lisbon, Portugal*

 Springer

A C.I.P. Catalogue record for this book is available from the Library of Congress.

ISBN-13 978-1-4419-3867-1
ISBN-10 0-387-26122-2 (e-book)
ISBN-13 978-0-387-26122-5 (e-book)

Published by Springer,
P.O. Box 17, 3300 AA Dordrecht, The Netherlands.

www.springeronline.com

Printed on acid-free paper

Printed in the Netherlands.

Dedication

This book is dedicated to

Our Wives

Contents

Preface

Integration of high-frequency analog filtering into the system Analog Front-End (AFE) is increasingly demanded for the ever growing high-speed communications and signal processing solutions with the corresponding advances in Integrated Circuit (IC) technology. Although the AFEs represent a small portion of the total mixed-signal system chip, they usually are its speed and performance bottleneck. Especially, the design of the AFEs becomes more and more challenging due to the continuous lowering of the supply and increasing of the operation speed, as well as noisying of the working environment driven by the constant growing digital signal processing (DSP) core.

This book presents a multirate sampled-data interpolation technique and its Switched-Capacitor (SC) implementation for very high frequency filtering (over hundreds of MHz) while having also dual inherent advantages of reducing the speed of the digital-to-analog converter and the DSP core together with the simplification of the post continuous-time smoothing filter.

The book is organized in eight chapters. This chapter presents an overview of the introductory aspects of the current state-of-the-art high-frequency SC filters and multirate filtering with emphasis on the SDA interpolation techniques for explicating the motivation and the objectives of the research work in this book.

Chapter 2 will describe the mathematical characterization of the conventional sampled-data analog interpolation with its input lower-rate S/H shaping distortion and will also introduce the ideal improved analog interpolation model with its traditional bi-phase SC structure implementation. Then, the development of the efficient multirate polyphase-based SC structures suitable for high-performance optimum-class improved analog interpolation filtering will be proposed. Different low-sensitivity circuit topologies with both Finite Impulse Response (FIR) and Infinite Impulse Response (IIR) characteristics will be developed, respectively, for low and high selectivity filtering.

Chapter 3 will present the practical IC technology imperfections related to IC implementation of SC multirate circuits that will be comprehensively investigated with respect to the power requirement issue, capacitance ratio mismatches, finite gain and bandwidth, input-referred DC offset sensitivity effects of the opamps, timing random-jitter and fixed periodic skew in the multirate clock phase generation as well as filter overall noise performance. All those practical design considerations are very useful in high-speed sampled-data analog integrated circuit design.

Chapter 4 will present advanced circuit techniques, i.e. gain- and offset-compensations, specialized for multirate SC filters and that are necessary to alleviate the imperfections of the analog integrated circuitry. Such techniques will be explored first for the basic building blocks: mismatch-free SC delay cells and SC accumulator, and later the impacts in the compensation of the overall system response will also be addressed and demonstrated through specific examples for both multirate FIR and IIR SC interpolating filters. Furthermore, the practical design trade-offs for utilization of such techniques will also be analyzed with respect to the accuracy versus speed and power.

Chapter 5 will set forth the design and implementation of a low-power SC baseband interpolating filter for NTSC/PAL digital video restitution system with CCIR-601 standards. The filter, which employs several novel optimized structures including coefficient-sharing, spread-reduction, semi-offset-compensation, mismatch-shaping, double-sampling and analog multirate/techniques, achieves a linear-phase lowpass response with 5.5-MHz bandwidth, 108 Msample/s output from 13.5 Msample/s video input. Both behavior-, transistor- and layout-extracted level simulations will be presented for illustrating the effectiveness of the circuit in 0.35 μm CMOS technology.

Chapter 6 will describe the design and implementation of a 2.5 V, 15-tap, 57 MHz SC FIR bandpass interpolating filter with 4-fold frequency up-translation for 22-24 MHz inputs at 80 MHz to 56-58 MHz outputs at 320MHz to be used in a Direct-Digital Frequency Synthesis (DDFS) system for wireless communication also in 0.35 μm CMOS. Special design considerations in both filter transfer function, circuit architectures, circuit building blocks as well as specific layout techniques for dealing with non-ideal properties in realization of the high-speed analog and digital clock circuits will be presented comprehensively in terms of the speed relaxation, noise and mismatching reduction.

Chapter 7 will then present the Printed-Circuit Board (PCB) design, experimental testing setup, as well as the measured results of the prototype interpolating filter chip built for the DDFS system described in Chapter 5. In addition to the measurement summary, a comparison among previously reported SC filters will also be offered.

Chapter 8 will finally draw the relevant concluding remarks.

Appendixes will be also provided for detailed mathematic derivation and analysis of the timing-skew errors in parallel sampled-data systems with S/H effects, namely, non-uniformly holding effects, and also the estimation scheme of the filter noise performance including opamp finite-gain and offset error analysis of SC building blocks.

Seng-Pan U, Ben
Rui Paulo Martins
José Epifânio da Franca

Acknowledgment

This work was developed under the support of the Research Committee of University of Macau, Integrated Circuits and Systems Group of Instituto Superior Técnico / Universidade Técnica de Lisboa, Fundação Oriente and Chipidea Microelectronics, S.A.. We also thank Terry Sai-Weng Sin for the assistance in formatting the text and figures as well as his contribution in timing-mismatch signal-to-noise mathematical analysis in Appendix 1. Finally, we would like to express enormous respect to our wifes for their constant understanding and endless support.

List of Abbreviations

AAF	:	Anti-Aliasing Filter
AC	:	Alternating Current
ADB	:	Active Delayed-Block
ADC	:	Analog-to-Digital Converter
AFE	:	Analog Front-End
AIF	:	Anti-Imaging Filters
AZ	:	Autozeroing
BPF	:	Band-Pass Filter
C-DFII	:	Complete Direct-Form II
CAD	:	Computer-Aided Design
CDMA	:	Code Division Multiple Access
CDS	:	Correlated-Double Sampling
CM	:	Common Mode
CMOS	:	Complementary Metal Oxide Semiconductor
CMFB	:	Common-Mode Feedback
CMRR	:	Common-Mode Rejection Ratio
CQFP	:	Ceramic Quad Flat-Pack
CT	:	Continuous-Time
DAC	:	Digital-to-Analog Converter
DB	:	Differentiator-Based
DC	:	Direct Current
DDFS	:	Direct-Digital Frequency Synthesis
DF	:	Direct-Form
DFII	:	Direct-Form II
DR	:	Dynamic Range

DSP	:	Digital Signal Processing
DT	:	Discrete-Time
DUT	:	Device Under Test
DVD	:	Digital Video Disks
EC	:	Error-storage Capacitor
EM	:	Electromagnetic
EMC	:	Electromagnetic Compatibility
ENBW	:	Equivalent Noise Bandwidth
ER	:	Extra Ripple
FFT	:	Fast Fourier Transform
FIR	:	Finite-Impulse-Response
GBW	:	Gain BandWidth
GOC	:	Gain- and Offset-Compensation
H-CDS	:	Holding Correlated-Double Sampling
IC	:	Integrated Circuit
IF	:	Intermediate-Frequency
IIR	:	Infinite Impulse Response
IM3	:	3rd-order Intermodulation Distortion
IN-CON	:	Input & Output timing-correlatively, Nonuniformly sampled & played out
IN-OU	:	Input Nonuniformly sampled, Output Uniformly played out
IS	:	Impulse-Sampled
IU-ON	:	Input Uniformly sampled, Output Nonuniformly played out
I-V	:	Current-to-Voltage
LC	:	Inductive-Capacitive
LPF	:	Low-Pass Filter
LVS	:	Layout versus Schematic
MF	:	Mismatch-Free
MCP-DFII	:	Mixed Cascade/Parallel Direct Form II
MOS	:	Metal-Oxide Semiconductor
MUX	:	Multiplexer
NTSC	:	National Television Standards Committee
OFR	:	Open-floating Resistor
OIP3	:	Output 3rd-order Intercept Point
OPAMP	:	operational amplifier
OTA	:	Operational Transconductance Amplifier
P-CDS	:	Predictive Correlated-Double Sampling

P-DFII	:	Parallel Direct Form II
PAL	:	Phase Alternation Line
PC	:	Parallel-Cyclic
PCB	:	Printed-Circuit Board
PCTSC	:	Parasitic-Compensated Toggle-Switched Capacitor
PM	:	Phase Margin
POG	:	Precise Opamp Gain
PSRR	:	Power Supply Rejection Ratio
PSS-AC	:	Periodic Swept Steady-State AC Analysis
QFP	:	Quad Flat-Pack
R-ADB	:	Recursive-ADB
RES	:	Rising-Edge Synchronizing
RF	:	Radio Frequency
ROM	:	Read-Only Memory
RUT	:	ROM Look-Up Table
SC	:	Switched-Capacitor
SDA	:	Sample-Data Analog
SDM	:	Sigma-Delta modulators
SDV	:	Switched Digital Video
SFDR	:	Spurious-Free Dynamic Range
S/H	:	Sample-and-Hold
SI	:	Switched-current
SMD	:	Surface-Mount Device
SINAD	:	Signal-to-Noise Plus Distortion Ratio
SNR	:	Signal-to-Noise Ratio
SSC	:	Same Sample Correction
T/H	:	Track-and-Hold
TDMA	:	Time Division Multiple Access
THD	:	Total Harmonic Distortion
TSC	:	Toggle-Switched Capacitor
TSI	:	Toggle-Switched Inverter
TV	:	Television
UC	:	UnCompensated
UGB	:	Unity-Gain Bandwidth
VCM	:	Common-Mode Voltage
VDSL	:	Video Digital Subscriber loop
V-I	:	Voltage-to-Current

List of Figures

List of Tables

Chapter 1

INTRODUCTION

1. HIGH-FREQUENCY INTEGRATED ANALOG FILTERING

Trends in high-speed communications and signal processing demand for the integration of high frequency analog filtering, traditionally implemented by external analog components, as much as possible on a system-chip to gain better performance and reliability at a reduced cost. Even considering that signal processing systems appear to be increasingly almost entirely digital, they still always necessitate to contain internally one or more integrated analog filtering functions or as their interface with the natural analog world. Moreover, filtering requirements at very high frequencies, where ultrafast sampling and digital circuitry with its associated data conversion, may not be realistic and economical, usually impose the use of analog techniques.

In general, the modern integrated analog filtering can be categorized in terms of implementation as Continuous-Time (CT), Discrete-Time (DT) and Sampled-Data Analog (SDA). Although CT filters especially like Gm-C [1.1, 1.2] and MOSFET-C [1.3, 1.4] have their superior capabilities of very high-frequency operation (up to hundred megahertz cutoff frequency), Switched-Capacitor (SC) filters, as the most dominant SDA structure, provide higher linearity and dynamic range with high accuracy and programmability of the time constants without requiring any complex tuning system as needed for CT filters [1.4, 1.5, 1.6, 1.7, 1.8, 1.9]. In addition, by taking advantage of the sampled-data processing nature, SC filters have also added superiority to the realization of the linear phase Finite-Impulse-Response (FIR) transfer function. Although SC filters still need CT front Anti-Aliasing filters (AAF) and post smoothing or Anti-Imaging filters (AIF), if multirate techniques are embedded in their structure, they will allow a significant relaxation of the CT front-end filtering.

Switched-current (SI) techniques could be used as another alternative for SDA filtering [1.10, 1.11, 1.12, 1.13, 1.14, 1.15, 1.16]. Although SI circuits operate in current-mode which has potentially lower voltage and wider bandwidth capability over the voltage mode, they need extra Voltage-Current (V-I) and Current-Voltage (I-V) conversion circuitries, and also, the attained precision as well as the dynamic range are both lower than those of SC filters [1.12, 1.16]. Moreover, a specific technique ally of SC circuits designated "Switched-Opamp" allows operation at very low voltage supply [1.17, 1.18, 1.19, 1.20, 1.21, 1.22, 1.23] with the state-of-the-art Switched-Opamp filter [1.20] and sigma-delta modulator [1.23] operating at 1 V and 0.7 V, respectively, while there is no similar technique available for SI circuits. Furthermore, and although SI circuits can be realized in low-cost standard digital CMOS technology, different techniques in the implementation of capacitors are also available for integrating a complete SC circuit chip in a digital CMOS process, e.g. metal-metal [1.24, 1.25], Fractal capacitor [1.26, 1.27], layer-sandwich [1.28, 1.29], Implant capacitor [1.30], Polysilicon-n-well [1.20, 1.22] and MOSFET-only [1.23, 1.31, 1.32].

High-frequency SC filters with over tens of MHz sampling rate have been emerging in different areas, namely in video signal processing [1.28, 1.33, 1.34, 1.35, 1.36, 1.37, 1.38, 1.39, 1.40, 1.41], magnetic disk read channels [1.42, 1.43], Switched Digital Video/Video Digital Subscriber loop (SDV/VDSL) [1.29], Intermediate-Frequency (IF) bandpass filtering [1.30, 1.44, 1.45, 1.46, 1.47], downconversion / subsampling with channel selection for wireless receivers [1.48, 1.49, 1.50, 1.51, 1.52], and many others [1.53, 1.54, 1.55, 1.56, 1.57]. Figure 1-1 presents previously reported high-frequency SC filters with output rate greater than 10 MHz in CMOS showing also their corresponding filter order. Among all, the highest order achieved is a 9-tap FIR function for a single-stage, 100 MHz to 33.3 MHz output, 3-path decimating filter [1.51], and also a 10th-order IIR non-optimum-class SC multirate filtering by cascading 5 biquads for 26 MHz to 13 MHz [1.52]. The highest output sampling rate achieved is 200 Ms/s reported by Severi et al. [1.56] in a double-sampling 2nd-order lowpass biquad session only. However, the ever growing area of high-speed data communication and processing obligates further development of SC filters by extending their operation to the hundreds of MHz range with even higher filter order using state-of-the-art CMOS technology under lower supply, and whose characteristics of speed and complexity must be targeted for the area beyond the trend lines of Figure 1-1.

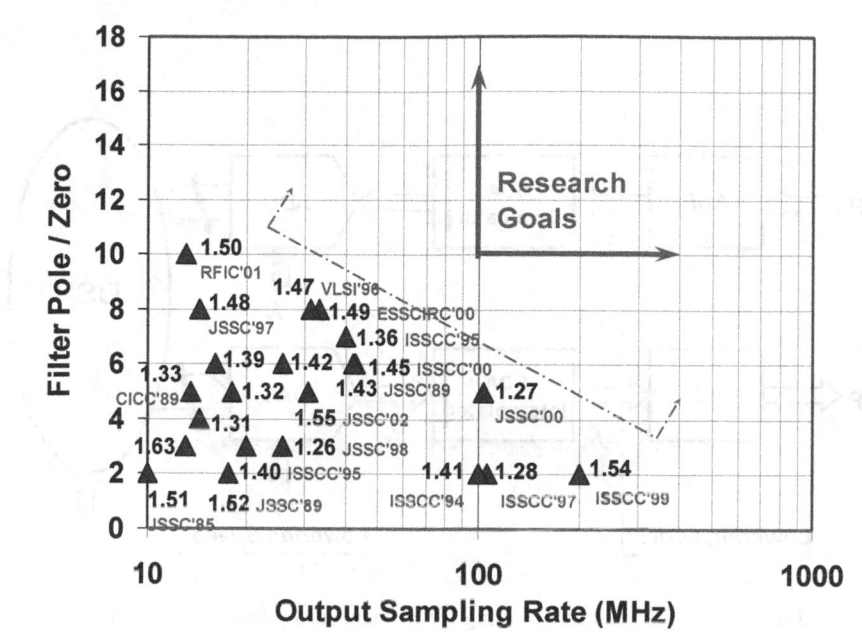

Figure 1-1. High-frequency Switched-Capacitor filters reported in CMOS

2. MULTIRATE SWITCHED-CAPACITOR CIRCUIT TECHNIQUES

The need for high-gain and bandwidth operational amplifiers (opamps) in standard SC circuits for high frequency of operation results in higher power consumption and also reduced design headroom. Therefore, to maximize the opamp bandwidth but still maintaining desired open-loop gain, different solutions like precise opamp gain (POG) [1.56] and pseudo-differential gain-enhancement replica amplifier [1.30] approaches have been proposed. The former solution needs to involve the precise opamp gain value as a parameter into circuit capacitor sizing for compensating the finite gain effects, thus, not only requiring an additional gain-control-closed-loop circuitry, to accurately steady the opamp gain, but also significantly increasing the design complexity especially for higher-order filter transfer function. Meanwhile, the gain enhancement in the latter depends on the mismatch between main- and replica-amplifiers, and the circuit has poor common-mode rejection ratio (CMRR), which is increasingly important for high-frequency mixed signal ICs.

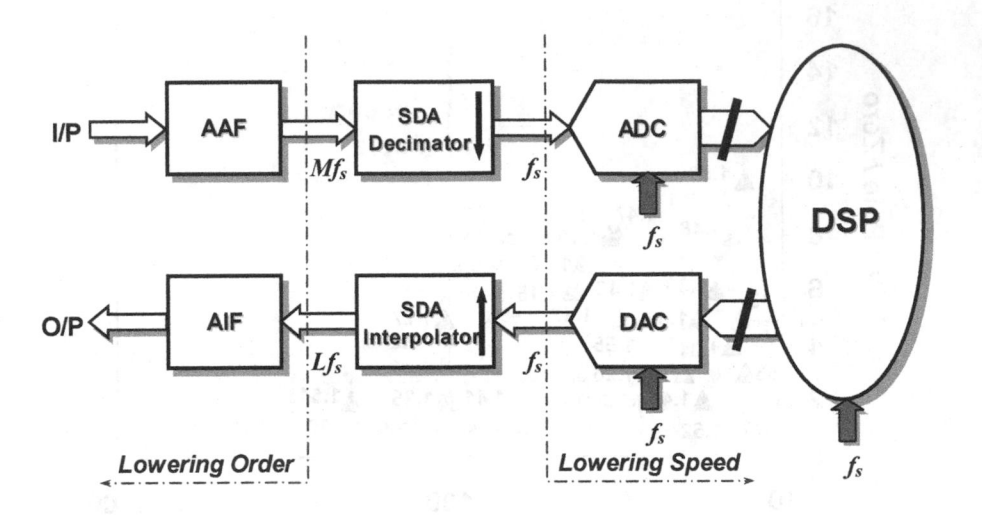

Figure 1-2. SDA multirate filtering for efficient analog front-end systems

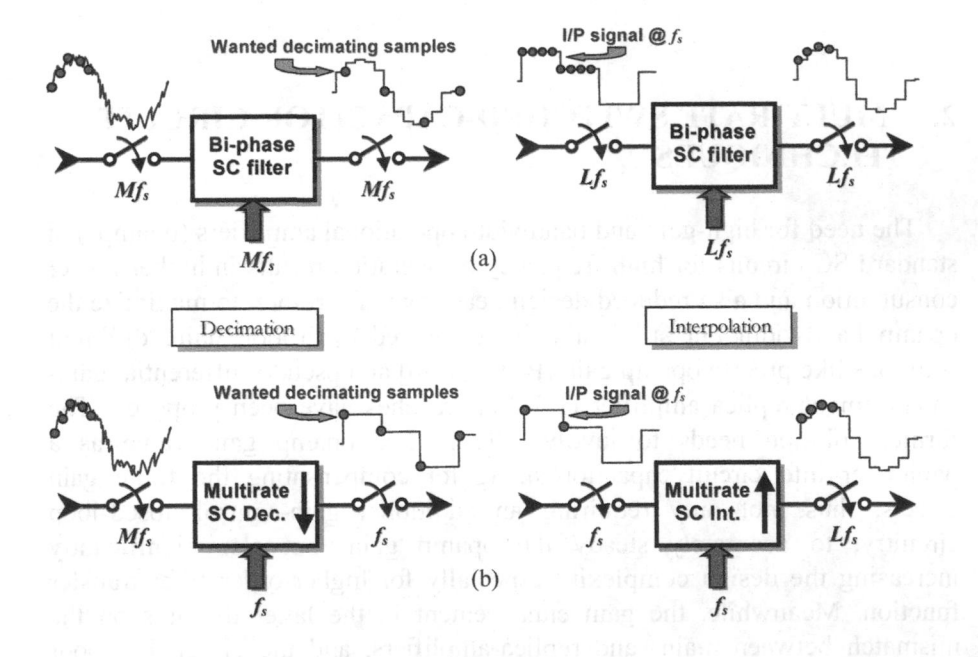

Figure 1-3. (a) Non-optimum-class and (b) Optimum-class decimation and interpolation filtering

On the other hand, it is also possible to relax the stringent speed requirement of the opamp through the enlargement of the circuit operating settling time by different specific system topologies, like double-sampling [1.29, 1.30, 1.34, 1.36, 1.42, 1.44, 1.45, 1.46, 1.51, 1.55, 1.56], parallel N-path [1.20, 1.28, 1.38, 1.47, 1.51, 1.58, 1.59] as well as multirate techniques [1.35, 1.37, 1.39, 1.40, 1.41, 1.43, 1.48, 1.49, 1.50, 1.51, 1.52, 1.57, 1.60]. Double-sampling is a simple and frequently-used approach for increasing the filter speed; however, it can only boost the operating speed twice, which is still not fast enough for very high-frequency and high-order applications. Parallel N-path structures are more suitable for narrow band applications, while they give rise to higher requirements of the anti-aliasing filter due to the multi-passband property within Nyquist, and suffer also from path mismatch effects that include fixed pattern noise (DC modulation) and in-band aliasing (signal image modulation).

By taking advantage of the inherent sampling rate conversion process, the multirate solution exhibits extra benefit by allowing not only a simplification in the CT anti-aliasing or anti-imaging filter but also, simultaneously, a further speed relaxation in data conversion and the power-hungry Digital Signal Processing (DSP) circuit core [1.61]. In Figure 1-2, the utilization of efficient SDA multirate filtering applied to an analog front-end system is presented. Multirate filtering, which includes decimators and interpolators corresponding to a discrete-time anti-aliasing and imaging-rejection filtering together with sampling rate reduction and increase, respectively, can be classified in 2 different implementations, i.e., non-optimum and optimum-class [1.62]. As shown in Figure 1-3, traditional Non-Optimum-Class designs use bi-phase SC filters operating at the highest sampling rate in the overall system, while in the opposite, the Optimum-Class realizations take advantage of the inherent multi-rate property and allow the opamps of the main filter core to operate, effectively, at the lowest sampling rate in the system, thus being especially appropriate for high-frequency filtering with added efficiency in power and silicon area as well as circuit design headroom.

3. SAMPLED-DATA INTERPOLATION TECHNIQUES

Various SC circuit structures for SDA decimating filters have been developed for use in high-frequency applications, such as video front-end [1.35, 1.38, 1.40, 1.41], magnetic disk read channels [1.43] and more recently downconversion/subsampling filtering for wireless receivers [1.48, 1.50, 1.51, 1.52]. On the other hand, SDA interpolation can be utilized in both baseband and frequency-translated modes which are presented in the

Figure 1-4(a) and (b), respectively. In these two modes, the interpolating filter corresponds either to the lowpass (baseband) or to the bandpass (frequency-translated) filtering associated to the sampling rate increase so as to attain dual benefits of lowering the speed of the DSP core and Digital-to-Analog Converter (DAC) as well as relaxing the post CT AIF filtering. Besides, the problematic glitch errors of DAC, especially for high-speed operation, will be eliminated due to the fact that the SDA interpolating filters will sample the settled signal at the DAC output.

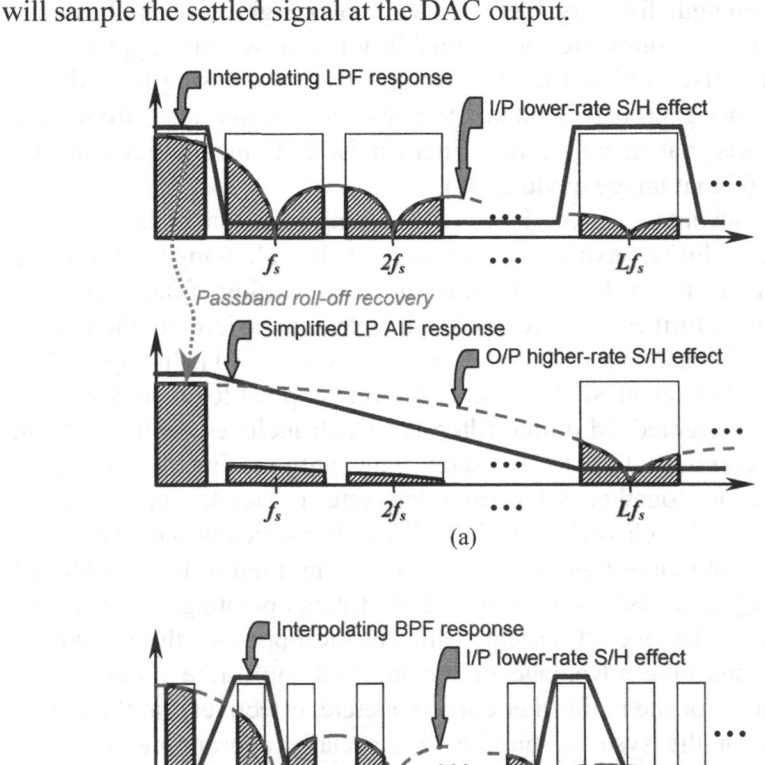

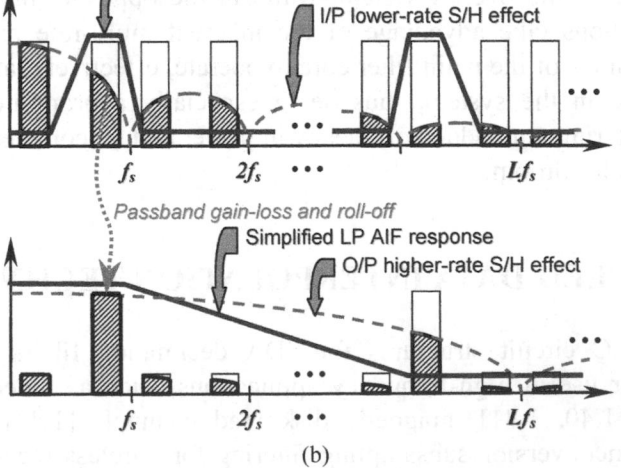

Figure 1-4. (a) Baseband (b) Frequency-translated interpolation filtering

It is worth to point out that the wanted signal band will be distorted by the inherent Sample-and-Hold (S/H) filtering shaping effect at the lower sampling rate from the DAC output. Especially for the wideband or the frequency-translated operation, the interested signal band will be seriously distorted by such shaping effect and will not be easy to recover through the traditional compensation either in the DSP or the CT reconstruction filter. Note that it is also possible to lower such distortion by adding zero-value samples to the DAC output signals in the amplifier (used either in between the DAC and CT reconstruction filtering or in the CT filter itself) together with a simultaneous increase of the CT filter passband gain. Nevertheless, it would not only force a very high slew-rate performance, which would in fact ultimately limit the speed of the circuit, but would also lead to more stringent demands on the bandwidth and gain of the active elements used in its construction, giving rise to limited design headroom and increased power consumption for very high-speed applications. Hence, the proper design approach to SDA interpolation must include mandatory immunity to such passband roll-off effect in practical implementations.

Several SC implementations have already involved the utilization of interpolating filters in different applications, e.g. video phone modem [1.63], PCM telephony [1.64], GSM baseband transmitter [1.65, 1.66]. However, the highest sampling rate achieved was only 13 MHz [1.65], and more importantly, all these circuits are implemented using non-optimum-class multirate structures, thus rendering not only large power and area consumption but also being unable to eliminate the undesired S/H shaping distortion at lower sampling rate from the DAC output.

Some specialized multirate SC structures for interpolating filters [1.67, 1.68, 1.69, 1.70, 1.71, 1.72] have also been investigated based on polyphase structures which are widely used in digital multirate signal processing for attaining extra computation efficiency [1.73, 1.74]. For finite-impulse response transfer functions, Direct-Form (DF) polyphase [1.67], Parallel-Cyclic (PC) polyphase [1.71] and Differentiator-Based (DB) non-recursive polyphase [1.70] SC interpolators have been proposed. However, the former DF and PC architectures are not practical for high selectivity filtering due to the resulting large number of SC branches and clock phases, which degrade the circuit performance with increased sensitivity to both capacitance ratios and switch timing. In addition, these three architectures cannot make good use of the inherent superiority of polyphase structures, i.e. low speed operation at input lower rate that can boost the filter speed while reducing the cost. For IIR transfer functions, SC interpolator building blocks combining either 1st- and 2nd-order building block or ladder-based recursive sessions together with DF polyphase networks have also been proposed [1.68, 1.69, 1.70], employing speed-non-optimum opamps in the filter core together with an output accumulator based either on a high-speed amplifier or on parasitic-sensitive unity-gain buffer. Furthermore, these specialized multirate SC interpolators require a more complicated design due to the need of modifying the original digital interpolating transfer function according with

$$H'(z) = H(z) \cdot \sum_{l=0}^{L-1} z^{-l} \qquad (1.1)$$

to account for the S/H shaping effect at the input lower rate [1.67].

4.   RESEARCH GOALS AND DESIGN CHALLENGES

The main objective of the research presented in this book is firstly to develop an efficient optimum-class of multirate SC structures suitable for higher-order (>10th-order) interpolating filters operating at very high frequency, i.e. in the order of hundreds of MHz output sampling rate in sub-micron CMOS technology, with supply voltage of 3 V or even lower at 2.5 V, to target in Figure 1-1 the high performance empty area (top-right). Moreover, such new structure must also eliminate the S/H shaping effects at the lower sampling rate.

In addition, the practical design challenges of real IC implementations, especially for very high-frequency operation, will also be investigated. Optimum-class decimating filters exhibit the highest-frequency signals at circuit input, so the handling of such signals can be basically done by passive element sampling, e.g. switches and capacitors; on the other hand, interpolating filters are required to generate high-frequency interpolated signals at the output, unavoidably by means of the active element, e.g. opamps. Hence, the design of a high-speed, high-linearity, mismatch-insensitive as well as low-power active output stage is still one of the most challenging tasks.

The filter order and especially the coefficient spread are normally proportional to the sampling rate increase factor, so, for some cases, the order and spread would be too large for practical IC implementation in terms of the speed, area and physical matching limitation. Therefore, a specific optimum design of the filter transfer function together with an elegant and efficient circuit structure would be mandatory.

Although the multirate structures are relatively less sensitive to the parallel path mismatch effects in the overall circuit than that in pure parallel

N-path structures, the mismatches caused by finite gain and offset as well as clock timing-skew especially in the last output stages will still degrade the system signal-to-noise tone ratio. Novel circuit structures insensitive to gain and offset mismatches are also important for high performance interpolation applications.

Clock generation becomes also an exceptional and vital part of multirate SC circuits due to the inherent multiple clock phase requirements. An efficient multiple-phase encoding logic, and more importantly, reduced phase skews must be considered both in the design systematic and process random variation. Furthermore, the digital coupling noise including dI/dt supply noise and substrate noise must be dealt with, due to the increased digital circuitry that is integrated nearby and in the same substrate of the sensitive analog circuitry.

The design of high-bandwidth and high-gain opamps with lower noise and power consumption as well as satisfactory linearity, in addition to the reduction of the charge injection and clock-feedthrough errors imposed by the enlarged switches and smaller capacitance, continue to be as always the most challenging tasks for very high-frequency SC circuits.

Any multirate or even a standard SC filter to operate in the hundreds of MHz range must address most of the above challenges, in terms of the choice of system architecture, circuit implementation and layout. Throughout this book, alternative approaches to tackle these challenges will be presented, and their impacts on the system overall performance will also be set forth.

With the proposed improvement techniques, IC prototypes based on the structures mentioned above will be implemented in a state-of-the-art sub-micron CMOS process. Such prototypes will target both baseband-mode and frequency-translated mode operations, corresponding to two of the most typical applications, i.e. the analog front-end filtering for a CCIR601 NTSC/PAL digital video system and the Direct-Digital Frequency Synthesis (DDFS) system for wireless communications. Finally, the experimental results will be provided to validate the referred circuit topologies and design methodologies.

REFERENCES

[1.1] N. Rao, V. Balan and R. Contreras, "A 3V 10-100-MHz Continuous-Time Seventh Order 0.05° Equiripple Linear Phase Filter", in *ISSCC Digest of Technical Papers*, pp. 44-46, Feb.1999.

[1.2] G.Bollati, S.Marchese, M.Demicheli, R.Castello, "An Eighth-Order CMOS Low-Pass Filter with 30–120 MHz Tuning Range and Programmable Boost," *IEEE J. Solid-State Circuits*, Vol.36, No.7, pp.1056-1066, Jul.2001.

[1.3] G. Groenewold, "Low-power MOSFET-C 120 MHz Bessel allpass filter with extended tuning range," *IEE Proc. Circuits, Devices and Sys.*, vol.147, no.1, pp. 28–34, Feb.2000.

[1.4] H. Khorramabadi and P. R. Gray, "High-frequency CMOS continuous-time filters," *IEEE J. Solid-State Circuits*, vol. SSC-19, pp.939–948, 1984.

[1.5] Y.P.Tsividis, "Integrated continuous-time filter design - An overview," *IEEE J. Solid-State Circuits*, vol.29, No.3, pp.166-176, Mar. 1994.

[1.6] R. Castello, F.Montecchi, F.Rezzi, A.Baschirotto, "Low-voltage analog filters," *IEEE Trans. on Circuits and Systems I: Fundamental Theory and Applications*, Vol.42, No.11, pp .827-840, Nov. 1995.

[1.7] R.Castello, I.Bietti, F.Svelto, "High-frequency filters in deep-submicron CMOS technology," in *ISSCC Digest of Technical Papers*, pp74-75, Feb.1999.

[1.8] José Moreira, *Design Techniques for Low-Power, High Dynamic Range Continuous-Time Filters*, Ph.D. Dissertation, Instituto Superior Técnico, Portugal, 1999.

[1.9] Y.P.Tsividis, "Continuous-time filters in telecommunications chips," *IEEE Communications Magazine*, pp.132-137, Apr. 2001.

[1.10] J.B.Hughes, N.C.Bird, I.C.Macbeth, "Switched-Currents – A new technique for analog sampled-data signal processing," in *Proc. IEEE International Symposium on Circuits and Systems (ISCAS)*, pp.1584-1587, May 1989.

[1.11] C.Toumazou, J.B.Hughes, N.C.Battersby, *Switched-Currents: an Analogue Technique for Digital Technology*, Peter Peregrinus Ltd, 1993.

[1.12] G.C.Temes, P.Deval, V.Valencia, "SC circuits: state of the art compared to SI techniques," in *Proc. IEEE International Symposium on Circuits and Systems (ISCAS)*, Vol.2, pp.1231-4, May 1993.

[1.13] J.B.Hughes, K.W.Moulding, "An 8MHz, 80Ms/s Switched-Current filter," in *ISSCC Digest of Technical Papers*, pp.60-61, Feb.1994.

[1.14] Y.L.Cheung, A.Buchwald, "A sampled-data Switched-Current analog 16-tap FIR filter with digitally programmable coefficients in 0.8 μm CMOS," in *ISSCC Digest of Technical Papers*, pp.54-55, Feb.1997.

[1.15] F.A.Farag, C.Galup-Montoro, M.C.Schneider, "Digitally programmable Switched-Current FIR filter for low-voltage applications," *IEEE J. Solid-State Circuits*, vol.35, No.4, pp.637-641, Apr. 2000.

[1.16] J.B.Hughesm A.Worapishet, C.Toumazou, "Switched-Capacitors versus Switched-Currents: a theoretical comparison," in *Proc. IEEE International Symposium on Circuits and Systems (ISCAS)*, Vol.II, pp.409-412, May 2000.

[1.17] J.Crols, M.Steyaert, "Switched-opamp: An approach to realize full CMOS Switched-Capacitor circuits at very low power supply voltages," *IEEE J. Solid-State Circuits*, Vol.29, pp.936-942, Aug. 1994.

[1.18] A.Baschirotto, R.Castello, "A 1 V 1.8 MHz CMOS switched-opamp SC filter with rail-to-rail output swing," in *ISSCC Digest Technical Papers*, pp.58-59, Feb.1997.

[1.19] V.Peluso, P.Vancorenland, A.Marques, M.Steyaert, W.Sansen, "A 900 mV 40 µW switched opamp ΔΣ modulator with 77 dB dynamic range," in *ISSCC Digest Technical Papers*, pp.68-69, Feb.1998

[1.20] V.S.L.Cheung, H.C.Luong, W.H.ki, "A 1 V CMOS Switched-Opamp Switched-capacitor pseudo-2-path filter," in *ISSCC Digest Technical Papers*, pp.154-155, Feb.2000.

[1.21] M. Waltari, K.A.I.Halonen, "1-V 9-bit pipelined switched-opamp ADC," *IEEE J. Solid-State Circuits*, Vol.36, pp.129-134, Jan. 2001.

[1.22] V.S.L.Cheung, H.C.Luong, W.H.ki, "A 1 V 10.7 MHz switched-opamp bandpass ΣΔ modulator using double-sampling finite-gain-compensation technique," in *ISSCC Digest Technical Papers*, pp.52-53, Feb.2001.

[1.23] J.Sauerbrey, T.Tille, D.Schnitt-Landsiedel, R.Thewes, "A 0.7V MOSFET-only switched-opamp ΣΔ modulator," in *ISSCC Digest Technical Papers*, pp.310-311, Feb.2002.

[1.24] L.A.Williams, "An audio DAC with 90 dB linearity using MOS to metal-metal charge transfer," in *ISSCC Digest Technical Papers*, pp.58-59, Feb.1998.

[1.25] E.Fogelman, I.Galton, W.Huff, H.Jensen, "A 3.3-V single-poly CMOS audio ADC delta-sigma modulator with 98-dB peak SINAD and 105-dB peak SFDR," *IEEE J. Solid-State Circuits*, Vol.35, pp.297-307, Mar. 2000.

[1.26] H.Samavati, A.Hajimiri, A.R.Shahani, G.N.Nasserbakht, T.H.Lee,"Fractal capacitors," *IEEE J. Solid-State Circuits*, Vol.33, pp.2035-2041, Dec. 1998.

[1.27] R.Aparicio, A.Hajimiri, "Capacity limits and matching properties of integrated capacitors," *IEEE J. Solid-State Circuits*, vol.37, pp.384-393, Mar. 2002.

[1.28] P.J.Quinn, "High-accuracy charge-redistribution SC video bandpass filter in standard CMOS," *IEEE J. Solid-State Circuits*, vol.33, No.7, pp.963-975, Jul.1998.

[1.29] U.K.Moon, "CMOS High-Frequency Switched-Capacitor filters for telecommunication applications," *IEEE J. Solid-State Circuits*, vol.35, No.2, pp.212-219, Feb. 2000.

[1.30] A.Nagari, G.Nicollini, "A 3 V 10 MHz pseudo-differential SC bandpass filter using gain enhancement replica amplifier," in *ISSCC Dig. Tech. Papers*, pp.52-53, Feb.1997.

[1.31] H.Yoshizawa, Y.Huang, P.F.Ferguson,G.C.Temes, "MOSFET-only switched-capacitor circuits in digital CMOS technology," *IEEE J. Solid-State Circuits*, Vol.34, pp.734-747, Jun. 1999.

[1.32] T.Tille, J.Sauerbrey, D.Schnitt-Landsiede,l "A 1.8-V MOSFET-only ΣΔ modulator using substrate biased depletion-mode MOS capacitors in series compensation," *IEEE J. Solid-State Circuits*, Vol.36, pp.1041-1047, Jul. 2001.

[1.33] K.Matsui, T.Matsuura, S.Fukasawa, Y.Izawa, Y.Toba, N.Miyake, K.Nagasawa, "CMOS video filters using Switched Capacitor 14-MHz circuits," *IEEE J. Solid-State Circuits*, Vol.SC-20, No.6, pp.1096-1102, Dec.1985.

[1.34] M.S.Tawfik, P.Senn, "A 3.6-MHz cutoff frequency CMOS Elliptic low-pass Switched-Capacitor ladder filter for video communication," *IEEE J. Solid-State Circuits*, Vol.SC-22, No.3, pp.378-384, Jun.1987.

[1.35] R.P.Martins, J.E.Franca, "A 2.4µm CMOS Switched-Capacitor video decimator with sampling rate reduction from 40.5MHz to 13.5MHz," in *Proc. IEEE Custom Integrated Circuits Conference (CICC)*, pp. 25.4/1 -25.4/4, May 1989.

[1.36] J.F.F.Rijns, H.Wallinga, "Spectral analysis of double-sampling Switched-Capacitor filters," *IEEE Trans. Circuits and Systems*, Vol.38, No.11, pp.1269-1279, Nov.1991.

[1.37] K.A.Nishimura, P.R.Gray, "A monolithic analog video comb filter in 1.2-µm CMOS," *IEEE J. Solid-State Circuits*, Vol.28, No.12, pp.1331-1339, Dec.1993.

[1.38] S.K.Berg, P.J.Hurst, S.H.Lewis, P.T.Wong, "A Switched-Capacitor filter in 2μm CMOS using parallelism to sample at 80MHz," in *ISSCC Dig. Tech. Papers*, pp.62-63, Feb.1994.

[1.39] S.Dosho, H.Kurimoto, M.Ozasa, T.Okamoto, N.Yanagisawa, N.Tamagawa, "A Comb filter with Switched-Capacitor delay lines for analog video processor," in *IEEE Symposium on VLSI Circuits Digest Technical Papers*, pp.54-55, Feb.1996.

[1.40] Ping Wang, J.E.Franca, "A CMOS 1.0-μm two-dimensional analog multirate system for real-time image processing," *IEEE J. Solid-State Circuits*, Vol.32, pp.1037-1048, Jul.1997.

[1.41] F.A.P.Barúqui, A.Petraglia, J.E.Franca, S.K.Mitra, "CMOS Switched-Capacitor decimation filter for mixed-signal video applications," in *Proc. European Solid-State Circuits Conference (ESSCIRC)*, Sep.2000.

[1.42] B.C.Rothenberg, S.H.Lewis, P.J.Hurst, "A 20Msample/s Switched-Capacitor finite-impulse-response filter in 2μm CMOS," in *ISSCC Digest Technical Papers*, pp.210-211, Feb.1995.

[1.43] G.T.Uehara, P.R.Gray, "A 100MHz output rate analog-to-digital interface for PRML magnetic-disk read channels in 1.2μm CMOS," in *ISSCC Digest Technical Papers*, pp.280-281, Feb.1994.

[1.44] B-S Soon, "Switched-Capacitor high-Q bandpass filter for IF application," *IEEE J. Solid-State Circuits*, Vol.SC-21, No.6, pp.924-933, Dec.1986.

[1.45] B-S Soon, "A 10.7 MHz Switched-Capacitor bandpass filter," *IEEE J. Solid-State Circuits*, Vol.24, pp.320-324, Apr.1989.

[1.46] A.Nagari, A.Baschirotto, F.Montecchi, R.Castello, "A 10.7-MHz BiCMOS high-Q double-sampled SC bandpass filter," *IEEE J. Solid-State Circuits*, Vol.32, pp.1491-1498, Oct.1997.

[1.47] K.V.Hartingsveldt, P.Quinn, A.V.Roermund, "A. 10.7MHz CMOS SC Radio IF Filter with Variable Gain and a Q of 55," in *ISSCC Digest Technical Papers*, pp.152-153, Feb.2000.

[1.48] D.H.Shen, C-M.Hwang, B.B.Lusignan, B.A.Wooley, "A 900-MHz Integrated Integrated Discrete-Time Filtering RF Front-End," in *ISSCC Digest Technical Papers*, pp.54-55, Feb.1996.

[1.49] T.B.Cho, G.Chien, F.Brianti, P.R.Gray, "A power-optimized CMOS baseband channel filter and ADC for cordless applications," in *IEEE Symposium on VLSI Circuits Digest Technical Papers*, pp.64-65, 1996.

[1.50] P.J.Chang, A.Rofougaran, A.A.Abidi, "A CMOS channel-select filter for a direct-conversion wireless receiver," *IEEE J. Solid-State Circuits*, Vol.32, pp.722-729, May 1997.

[1.51] R.F.Neves, J.E.Franca, "A CMOS Switched-Capacitor bandpass filter with 100 MSample/s input sampling and frequency downconversion," in *Proc. European Solid-State Circuits Conference (ESSCIRC)*, pp. 248-251, Sep.2000.

[1.52] Yi-Huei Chen, Jenn-Chyou Bor; Po-Chiun Huang, "A 2.5 V CMOS Switched-Capacitor channel-select filter with image rejection and automatic gain control," in *IEEE Radio Frequency Integrated Circuits (RFIC) Symposium Digest of Papers*, pp.111-114, 2001.

[1.53] D.B.Ribner, M.A.Copeland, "Biquad Alternative for High-Frequency Switched-Capacitor Filters," *IEEE J. Solid-State Circuits*, Vol.SC-20, No.6, pp.1085-1094, Dec.1985.

[1.54] G.Nicollini, F.Moretti, M.Conti, "High-frequency fully differential filter using operational amplifiers without common-mode feedback," *IEEE J. Solid-State Circuits*, vol.24, No.3, pp.803-813, Jun. 1989.

[1.55] A.Baschirotto, F.Montecchi, R.Castello, "A 150 Msample/s 20 mW BiCMOS switched-capacitor biquad using precise gain op amps," in *ISSCC Digest Technical Papers*, pp.212-213, Feb.1995.

[1.56] F.Severi, A.Baschirotto, R.Castello, "A 200Msample/s 10mW Switched-Capacitor Filter in 0.5µm CMOS Technology" in *ISSCC Digest Technical Papers*, pp.400-401, Feb.1999.

[1.57] S.Azuma, S.Kawama, K.Iizuka, M.Miyamoto, D.Senderowicz, "Embedded Anti-Aliasing in Switched-Capacitor Ladder Filters with variable gain and offset compensation," *IEEE J. Solid-State Circuits*, vol.37, No.3, pp.349-356, Mar. 2002.

[1.58] M.B.Ghaderi, J.A.Nossek, G.C.Temes, "Narrow-band Switched-Capacitor bandpass filters," *IEEE Trans. Circuits and Systems*, Vol.CAS-8, pp.557-571, Aug.1982.

[1.59] D.C.von Grunigen, R.P.Sigg, J.Schmid, G.S.Moschytz, H.Melchior, "An integrated CMOS Switched-Capacitor bandpass filter based on N-Path and frequency-sampling principles," *IEEE J. Solid-State Circuits*, Vol.SC-18, pp.753-761, Dec.1983.

[1.60] R.P.Martins, J.E.Franca, F.Maloberti, "An optimum CMOS Switched-Capacitor antialiasing decimating filter," *IEEE J. Solid-State Circuits*, Vol.28 No.9, pp.962-970, Sep. 1993.

[1.61] J.E.Franca, A.Petraglia, S.K.Mitra, "Multirate analog-digital systems for signal processing and conversion," *Proc. of The IEEE*, Vol.85, No.2, pp.242-262, Feb.1997.

[1.62] J.E.Franca, R.P.Martins, "IIR Switched-Capacitor decimator building blocks with optimum implementation," *IEEE Trans. Circuits and Systems*, Vol. CAS-37, No.1, pp.81-90, Jan. 1990.

[1.63] C.W.Solomon, L.Ozcolak, G.Sellani, W.E.Brisco, "CMOS analog front-end for conversational video phone modem," in *Proc. IEEE Custom Integrated Circuits Conference (CICC)*, pp.7.4/1-7.4/5, 1989.

[1.64] D.Senderowicz, G.Nicollini, P.Confalonieri, C.Crippa, C.Dallavalle, "PCM Telephony: Reduced architecture for a D/A converter and filter combination," *IEEE J. Solid-State Circuits*, vol. 25, pp.987–995, Aug.1990.

[1.65] B.Baggini, L.Coppero, G.Gazzoli, L.Sforzini, F.Maloberti, G.Palmisano, "Integrated digital modulator and analog front-end for GSM digital cellular mobile radio system," *Proc. IEEE Custom Integrated Circuits Conference (CICC)*, pp.7.6/1 -7.6/4, 1991.

[1.66] C.S.Wong, "A 3-V GSM baseband transmitter," *IEEE J. Solid-State Circuits*, vol.34, No.5, pp.725-730, May 1999.

[1.67] J.E.Franca, "Non-recursive polyphase Switched-Capacitor decimators and interpolators," *IEEE Trans. Circuits and Systems*, Vol. CAS-32, pp. 877-887, Sep.1985.

[1.68] R.P.Martins, J.E.Franca, "Infinite impulse response Switched-Capacitor interpolators with optimum implementation", in *Proc. IEEE International Symposium on Circuits and Systems (ISCAS)*, pp.2193-2196, May 1990.

[1.69] R.P.Martins, J.E.Franca, "Novel second-order Switched-Capacitor interpolator", *Electronics Letters*, Vol.28 No.2, pp.348-350, Feb.1992.

[1.70] C.-Y.Wu, S.Y.Huang, T.-C.Yu, Y.-Y.Shieu, "Non-recursive Switched-Capacitor decimator and interpolator circuits," in *Proc. IEEE International Symposium on Circuits and Systems (ISCAS)*, pp.1215-1218, 1992.

[1.71] K.Kato, T.Kikui, Y.Hirata, T.Matsumoto, T.Takebe, "SC FIR interpolation filters using parallel cyclic networks," in *Proc. IEEE International Symposium on Circuits and Systems (ISCAS)*, pp.723-726, 1994.

[1.72] P.J.Santos, J.E.Franca, "Switched-capacitor interpolator for direct-digital frequency synthesizers," in *Proc. IEEE International Symposium on Circuits and Systems (ISCAS)*, Vol.2, pp.228 -231, 1998.

[1.73] R.E.Crochiere, L.R.Rabiner, *Multirate Digital Signal Processing*, Prentice-Hall, Inc., NJ, 1983.

[1.74] S.K.Mitra, J.F.Kaiser, *Handbook for Digital Signal Processing*, John Wiley & Sons, Inc., 1993.

Chapter 2

IMPROVED MULTIRATE POLYPHASE-BASED INTERPOLATION STRUCTURES

1. INTRODUCTION

The design of improved SC structures for interpolating filtering embraces first the speed relaxation and number reduction of the opamps in the circuit for the optimum-class multirate realization, and secondly the elimination of the input lower-rate S/H shaping effect which then leads the SDA interpolation to operate in a similar manner as its digital counterpart. Previously available SC interpolator structures cannot fulfill all the above requirements [2.1, 2.2, 2.3, 2.4, 2.5, 2.6, 2.7, 2.8, 2.9, 2.10].

This chapter will first characterize the conventional sampled-data analog interpolation with its input lower-rate S/H shaping distortion and propose the ideal improved analog interpolation model and its traditional bi-phase SC structure. Then, by proving first the effectiveness of the employment of multirate polyphase structure for optimum-class improved analog interpolation that will completely get rid of the input lower-rate S/H shaping effect in the entire frequency axis, a family of multirate SC structures with increased speed, power and silicon area efficiency for IC realizations, namely, Active Delayed-Block (ADB) polyphase-based structures, will be proposed by combining the novel input sampling technique and the Direct-Form (DF) polyphase structures with original digital prototype interpolation filtering transfer function [2.11, 2.12, 2.13]. Two types of SC structure will be presented one employing a novel L-output-accumulator suitable for high-frequency operation, and the other using a one-output-accumulator yielding a reduced component count. Both canonic- and non-canonic-forms ADB polyphase structures with respect to the required actual delay terms for

ADB's will be proposed for both higher-order linear-phase FIR and high-selectivity/wideband IIR interpolation functions. Moreover, the specific low-sensitivity IIR multirate structures will also be investigated for higher-order filtering.

2.   CONVENTIONAL AND IMPROVED ANALOG INTERPOLATION

Interpolation by a factor L corresponds to the process of sampling rate increase from f_s to Lf_s. Pure digital implementation of interpolation comprehends the combined operation of an up-sampler, for increasing the sampling rate from f_s to Lf_s and inserting (L-1) zero-valued samples between two consecutive input samples, and an interpolation filter for removing the unwanted frequency-translated image components associated with the signal sampled at the input lower rate. The spectrum of the resulting ideal output interpolated samples $x_{it}[nT_{it}]$ is given by

$$X_{it}(e^{j\omega}) = X_e(e^{j\omega}) \cdot H(e^{j\omega}) = X(e^{j\omega L}) \cdot H(e^{j\omega}), \quad \omega = \Omega T_{it} = \Omega T_s / L \quad (2.1)$$

where $X_e(e^{j\omega})$ and $X(e^{j\omega L})$, are, respectively, the spectrum of the up-sampled and the original samples, and $H(e^{j\omega})$ is the ideal frequency response of the interpolation filter (gain = L, cutoff frequency $\omega_c = \pi /L$). Therefore, an ideal S/H interpolated output signal $x_{it}(t)$ can be obtained by passing such interpolated samples through an ideal hold circuit, and its spectrum is represented by

$$X_{it}(j\Omega) = X_{it}(e^{j\Omega T_{it}}) \cdot \frac{T_{it} \sin(\Omega T_{it}/2)}{\Omega T_{it}/2} e^{-j\Omega T_{it}/2} \quad (2.2)$$

In the (sampled-data) analog case, the exact interpolation (as in the above digital case) is not possible due to the input S/H signal, then it must be described by the conventional analog interpolation model in Figure 2-1(a). The analog interpolating filter, which can be analyzed as a discrete-time processor operating at Lf_s with an output hold at Lf_s, will sample and process the input signal at Lf_s (thus having L successive equal-value samples owing to the constant-held input within a full sampling period $1/f_s$) and its operation is depicted in Figure 2-1(b), both in time and frequency domains. The spectrum of the input S/H samples $x'_e[nT_{it}]$ can be expressed in terms of the spectrum of the up-sampled discrete samples $x_e[nT_{it}]$ by

$$X'_e(e^{j\omega}) = X_e(e^{j\omega}) \cdot H^0_{SD}(e^{j\omega}) \quad (2.3)$$

where

$$H_{SD}^{0}(e^{j\omega}) = \frac{\sin(\omega L/2)}{\sin(\omega/2)} \cdot e^{-j(L-1)\frac{\omega}{2}}, \qquad \omega = \Omega T_{it} \qquad (2.4)$$

in which $\left| H_{SD}^{0}(e^{j\omega}) \right| = L$, for $\omega = 0$.

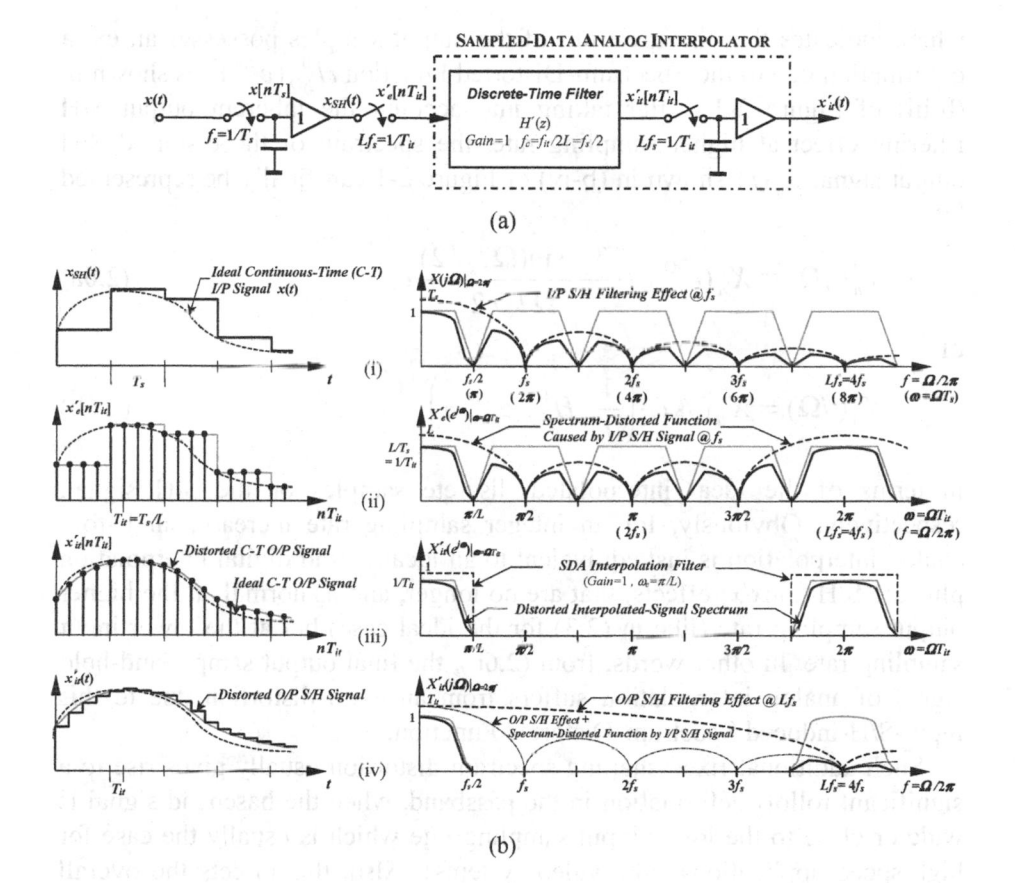

Figure 2-1. Conventional analog L-fold interpolation (a) Architecture model (b) Time- and frequency-domain illustration

From (2.4), the spectrum of the processed samples in an analog interpolation, as illustrated in (b-ii) of Figure 2-1, is a deformed version of $X_{e}(e^{j\omega})$ due to the multiplication by $H_{SD}^{0}(e^{j\omega})$ which is referred as Spectrum-Distorted function with a DC gain of L caused by the sampling of the constant-held input. Thus, a unity-gain interpolation filter ($H'(e^{j\omega}) = H(e^{j\omega})/L$) must be employed to process such samples, and the spectrum of the resulting output interpolated samples $x'_{it}[nT_{it}]$ is expressed as

$$X'_{it}(e^{j\omega}) = X'_e(e^{j\omega}) \cdot H'(e^{j\omega})$$
$$= X_e(e^{j\omega}) \cdot H^0_{SD}(e^{j\omega}) \cdot H'(e^{j\omega}), \qquad \omega = \Omega T_{it} \qquad (2.5)$$

which indicates that the spectrum of the output samples possesses an extra deformation due to the Spectrum-Distorted Function $H^0_{SD}(e^{j\omega})$, as shown in (b-iii) of Figure 2-1. After taking into account the inherent output S/H filtering effect at higher sampling rate, the spectrum of the distorted S/H output signal $x'_{it}(t)$ shown in (b-iv) of Figure 2-1 can finally be represented by

$$X'_{it}(j\Omega) = X_{it}(e^{j\Omega T_{it}}) \cdot \frac{T_s \cdot \sin(\Omega T_s / 2)}{\Omega T_s / 2} e^{-j\frac{\Omega T_s}{2}} \qquad (2.6a)$$

or

$$X'_{it}(j\Omega) = X_{it}(j\Omega) \cdot \left(\frac{1}{L} \cdot H^0_{SD}(e^{j\Omega T_{it}}) \right) \qquad (2.6b)$$

in terms of the ideal interpolated discrete samples or the S/H signal, respectively. Obviously, for an integer sampling rate increase, an L-fold analog interpolation is just equivalent to an ideal L-fold digital interpolation plus the S/H ($\sin x/x$) effects, that are no longer, and as normal, at the higher output sampling rate (like in (2.3) for the ideal case) but at the lower input sampling rate. In other words, from (2.6b), the final output sample-and-held signal of analog interpolation suffers from an extra distortion due to this input-S/H-induced Spectrum-Distorted Function.

Such additional fixed-shaping spectrum distortion usually gives rise to a significant rolloff deformation in the passband, when the baseband signal is wide or close to the lower input sampling rate which is usually the case for high-speed applications (like video systems). Also, this affects the overall system response when frequency-translated bandpass processing is required (like subsampling in wireless communications). Hence, an improved analog interpolation is presented in Figure 2-2(a) destined to eliminate such frequency shaping distortion, thus leading to an increased simplification and freedom in the design of both the passband and the stopband. Although the input signal is still sampled-and-held at lower rate, the ideal overall interpolation performance will be exactly equivalent to a digital interpolation, apart from the S/H effect at the higher sampling rate that is always present in sampled-data analog systems.

A simple SC implementation of this improved analog interpolation is illustrated in Figure 2-2(b) which combines a bi-phase SC filter operating at Lf_s with a special sampling by a front two-switch input interface that operates as an up-sampler by forcing the circuit input to connect to ground at the appropriate time, thus generating zero-valued samples. However, this approach belongs clearly to the non-optimum-class of implementation since the filter core needs to operate at the highest sampling rate of the overall system, and also, an additional DC gain (with value L) is necessary in the filter thus rendering inefficient coefficient spread which leads to large power and area consumption.

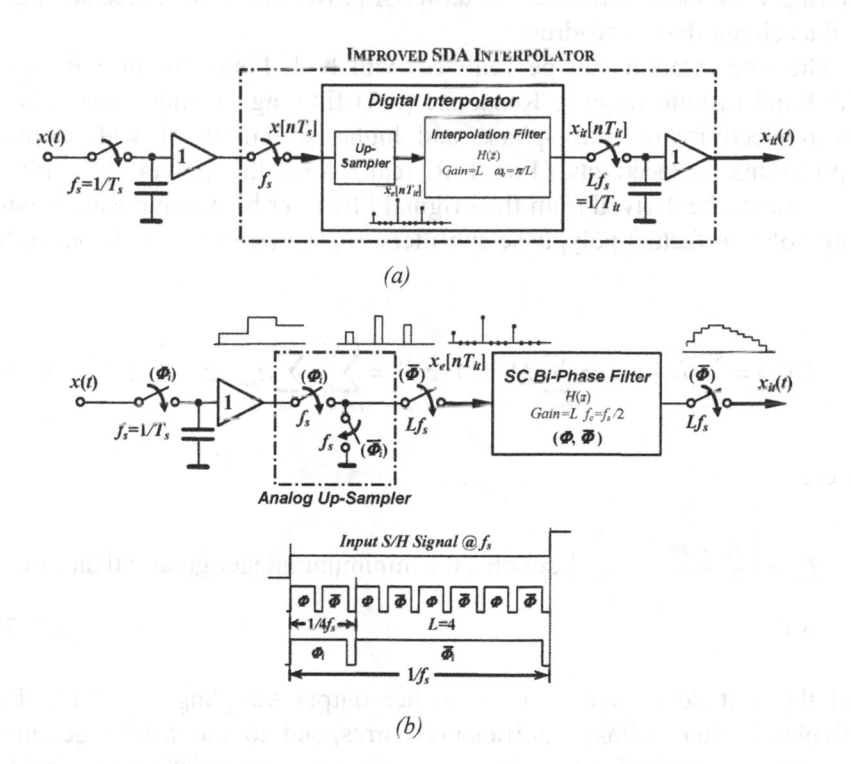

Figure 2-2. Improved Analog interpolation with reduced S/H effects   (a) Architecture Model   (b) Non-optimum SC implementation with a high-rate Bi-Phase filter

3. POLYPHASE STRUCTURES FOR OPTIMUM-CLASS IMPROVED ANALOG INTERPOLATION

An optimum-class realization of the improved analog interpolation, without lower input-rate S/H shaping distortion, can be achieved by employing polyphase decomposition which is an efficient and straightforward structure utilized in digital multirate filters [2.14, 2.15, 2.16]. Such realization, based on the original digital prototype interpolating transfer function without any modification, takes advantage of the inherent multirate property and allows the main filter core to operate, effectively, at the lowest sampling rate of the system, thus being appropriate for high-frequency filtering with added efficiency in terms of power and silicon area savings as well as circuit design headroom.

The interpolation can be realized with both Finite Impulse Response (FIR) and Infinite Impulse Response (IIR) filtering functions normally for lower-selectivity or linear-phase and higher-selectivity or wide-stopband applications, respectively. For FIR realization, the polyphase multirate structure can be derived from the original FIR filter by decomposing it into L (interpolation factor) polyphase subfilters $H_m(z)$ ($m=0,1,...,L-1$), according to

$$H(z) = \sum_{n-0}^{N-1} h_n \cdot z^{-n} = \sum_{m-0}^{L-1} H_m(z) \cdot z^{-m} = \sum_{m=0}^{L-1} \left(\sum_{i=0}^{I_m-1} h_{m+iL} z^{-iL} \right) \cdot z^{-m} \qquad (2.7a)$$

where

$$I_m = \left\lfloor \frac{N-m}{L} \right\rfloor$$ ($\lfloor x \rfloor$ denotes the minimum integer greater than or equal

to x) $\qquad (2.7b)$

and the unit delay refers to the higher output sampling rate $1/Lf_s$. Each polyphase filter, whose coefficients correspond to the L-fold decimated versions of original filter impulse response, approximates an all-pass function and each value of m corresponds to a different phase shift network. Hence, they all efficiently operate at input lower sampling rate and contribute with one nonzero output for each, which corresponds to one of the L outputs of the interpolating filter generated in a sweep mode from the zero[th] to the Lth polyphase filter by an output counter-clockwise commutator (at output higher rate) for each input sample [2.11, 2.12, 2.13].

Each polyphase filter can be simply implemented by Direct-Form (DF), thus being designated as *DF polyphase* structure. For simplicity, this is illustrated in Figure 2-3 with an example that demonstrates the effectiveness of the polyphase structure to achieve optimum-class improved interpolation filtering. Supposing that an input signal is required to be 2-fold interpolated with respect to a simple 3-tap FIR function, then the original digital transfer function of the interpolation filter is decomposed into a set of L polyphase filters $\{H_m(z),\ m=0,\ 1\}$, leading to the resulting polyphase structure of Figure 2-3, where all polyphase filters are realized with a DF structure.

The first polyphase filter produces an output sample given by

$$x_{it}\left[nT_{it}\right]=h_0 \cdot x\left[nT_{it}\right]+h_2 \cdot x\left[(n-2)T_{it}\right] \tag{2.8a}$$

which is equivalent to multiply the coefficient h_1 by a zero-valued sample. Similarly, since the second polyphase filter produces an output sample given by

$$x_{it}\left[(n+1)T_{it}\right]=h_1 \cdot x\left[(n+1)T_{it}\right] \tag{2.8b}$$

where $\quad x\left[nT_{it}\right]=x\left[(n+1)T_{it}\right]$

it is also equivalent to multiplying by zero the coefficients h_0 and h_2. Thus, such operation is equivalent to a digital interpolation where its zero-valued samples need to be created by a digital up-sampler.

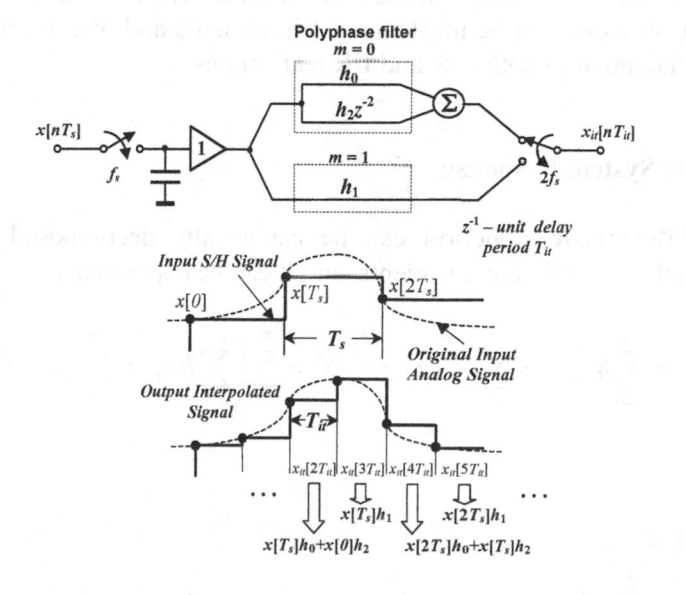

Figure 2-3. Improved analog interpolation with Optimum-class realization by Direct-Form polyphase structure (L=2)

In general, it is concluded that the DF polyphase interpolation, with original digital prototype transfer function, implements an improved analog interpolation without the input S/H filtering effect. Since every polyphase filter inherently operates at the lower input sampling rate, the input held signal is only sampled by the interpolator once per period. This also explains why the input S/H effect don't affect the overall system response of the polyphase-structure-based interpolation.

4.   MULTIRATE ADB POLYPHASE STRUCTURES

4.1   Canonic and Non-Canonic ADB Realizations

DF polyphase structure is appropriate only when the FIR filter length N is not much greater than the interpolation factor L, e.g. $N \leq 2L$, since it leads to circuits having a rather large number of time-interleaved SC branches and switching phases, which increase not only its complexity beyond practical acceptable limits but the sensitivity to mismatch of capacitance ratios and switch timing. Hence, a more general architecture, designated by Active-Delayed Block (ADB) was introduced [2.17, 2.18] that is a polyphase-based structure to overcome such limitations for filter length $N>2L$. Such ADB polyphase structure can be implemented in Canonic and Non-Canonic form with easy adaption to both FIR and IIR realizations.

4.1.1   FIR System Response

The FIR transfer function can be canonically decomposed in B_c+1 blocks, each with only L coefficients, and it can be expressed as

$$H(z) = \sum_{n=0}^{N-1} h_n z^{-n} = \sum_{b=0}^{B_c} G_b(z) \cdot (z^{-L})^b = \sum_{b=0}^{B_c} \left(\sum_{n=0}^{L-1} h_{n+bL} z^{-n} \right) \cdot (z^{-L})^b \quad (2.9a)$$

where

$$B_c = \left\lfloor \frac{N-L}{L} \right\rfloor, \qquad\qquad\qquad\qquad\qquad (2.9b)$$

The elements in each block b will have at least b delay terms z^{-L} (except $b=0$) that will be implemented by an SC ADB. Since each block $G_b(z)$ containing L coefficients (except the last block, $b=B_c$, which contains only $N-B_cL$ terms) can be decomposed again in a polyphase subfilter, that can be realized in DF structure with the sharing of a low speed serial ADB delay line composed by regular z^{-L} units, this structure is designated as *Canonic ADB Polyphase* structure [2.18].

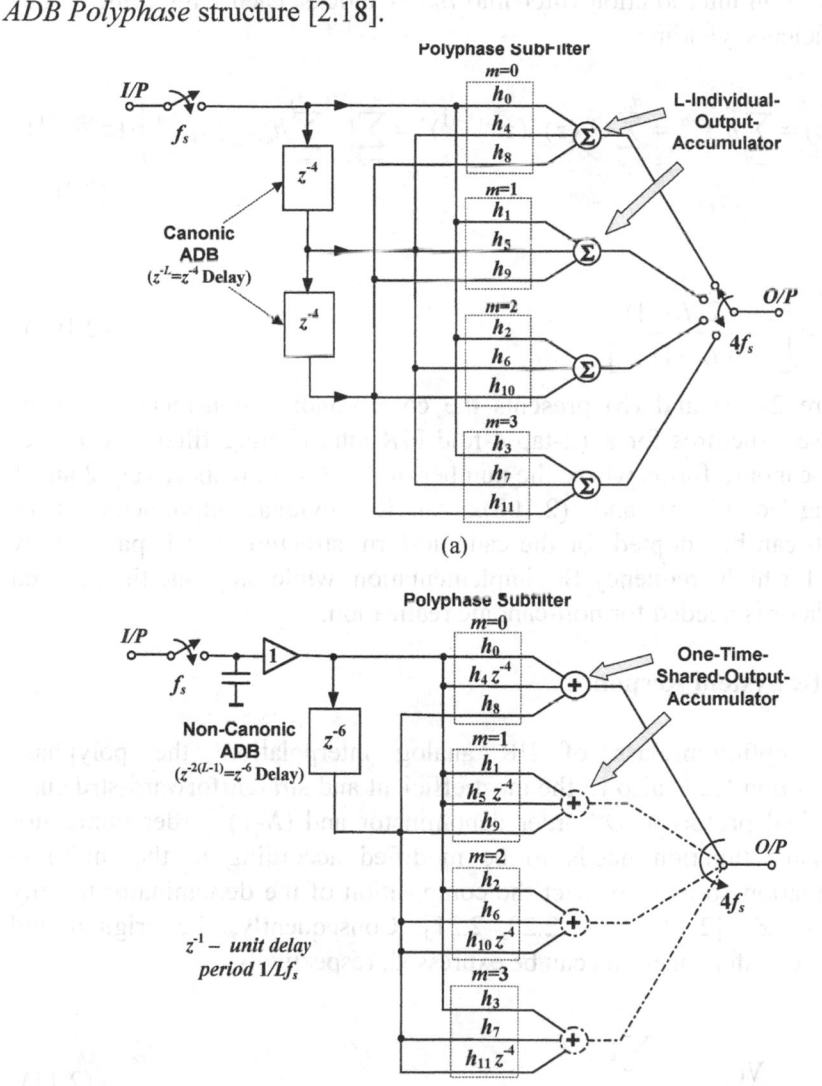

Figure 2-4. (a) Canonic-form (b) Non-canonic-form ADB polyphase structures for improved 4-fold 12-tap FIR interpolator

Minimizing the number of opamps in the ADB-based architecture can be achieved by reducing the number of both ADB's and accumulators, this can be obtained by decomposing the transfer function into blocks – $G_b(z)$ with more-than-L coefficients while making their shared delays larger than regular unit z^{-L}. Such realization is referred to as *Non-Canonic ADB Polyphase* structure, and can be obtained by decomposing the transfer function of an interpolation filter into $B_{nc}+1$ blocks, each with at most $2(L-1)$ coefficients, yielding

$$H(z) = \sum_{n=0}^{N-1} h_n z^{-n} = \sum_{b=0}^{B_{nc}} G_b(z) \cdot (z^{-2(L-1)})^b = \sum_{b=0}^{B_{nc}} \left(\sum_{n=0}^{2(L-1)-1} h_{n+b \cdot 2(L-1)} z^{-n} \right) \cdot (z^{-2(L-1)})^b$$

(2.10a)

where

$$B_{nc} = \left\lfloor \frac{N - 2(L-1)}{2(L-1)} \right\rfloor$$

(2.10b)

Figure 2-4(a) and (b) presents the corresponding non-recursive ADB polyphase structures for a 12-tap 4-fold FIR interpolating filter in canonic- and non-canonic-forms where the number of ADB's are respectively 2 and 1 according to (2.9b) and (2.10b). An L-individual-output-accumulator approach can be adopted for the canonic-form structure that is particularly suitable for high-frequency SC implementation, while only one time-shared accumulator is needed for non-canonic realization.

4.1.2 IIR System Response

For the optimum-class of IIR analog interpolation, the polyphase decomposition leads also to the most efficient and straightforward structure. The original prototype D^{th}-order denominator and $(N-1)^{th}$-order numerator IIR transfer function needs to be modified according to the multirate transformation so as to restrict the composition of the denominator to only powers of z^{-L} [2.14, 2.19, 2.20, 2.21]. Consequently, the original and modified transfer functions can be expressed, respectively, as

$$H(z) = \frac{N(z)}{D(z)} = \frac{\sum_{i=0}^{N-1} a_i z^{-i}}{1 - \sum_{j=1}^{D} b_j z^{-j}}$$

(2.11)

and

$$\hat{H}(z) = \frac{\hat{N}(z)}{\hat{D}(z)} = \frac{\displaystyle\sum_{i=0}^{(N-1)+D(L-1)} A_i z^{-i}}{1 - \displaystyle\sum_{j=1}^{D} B_{jL}(z^{-L})^j} \tag{2.12}$$

The particular form of (2.11), which allows the recursive part to operate at the lower input sampling rate, can be constructed by combining a non-recursive ADB polyphase structure together with a recursive Direct-Form II (DFII) structure for realizing, respectively, the numerator and the denominator polynomials. Such architecture, where the common delay blocks z^{-L} are realized by a low speed ADB serial delay line and are efficiently shared by both recursive and non-recursive parts, can be referred to as *Recursive-ADB (R-ADB) Polyphase* structure [2.21]. This offers a more general, straightforward and flexible design with enhanced efficiency in terms of amplifier speed and number of phases when compared with previous structures [2.6, 2.7, 2.10].

Like in the FIR counterpart, this R-ADB polyphase structure can also be implemented in canonic and non-canonic forms categorized by the corresponding delay of the shared ADB's. The former has L unit delays whereas the latter requires delays of $2(L-1)$ (except the 1st block that has always a unity delay). Thus, for a general case ($D \neq N-1$), the IIR modified transfer function in canonic form, which requires $\max(B_{cn}, B_{cd})$ SC ADB's can be reformulated as

$$\hat{H}(z) = \frac{\displaystyle\sum_{j=0}^{B_{cn}} \left(\sum_{i=0}^{L-1} A_{i+jL} z^{-i} \right) \cdot (z^{-L})^j}{1 - \displaystyle\sum_{j=1}^{B_{cd}} B_{jL}(z^{-L})^j} \tag{2.13a}$$

where

$$B_{cn} = \left\lfloor \frac{N + D(L-1) - L}{L} \right\rfloor \quad \& \quad B_{cd} = D \tag{2.13b}$$

while the non-canonic transfer function that requires $\max(B_{ncn}, B_{ncd})$ ADB's can be expressed as

$$\hat{H}(z) = \frac{A_0 + z^{-1} \cdot \displaystyle\sum_{j=0}^{B_{ncn}} \left(\sum_{i=0}^{2(L-1)-1} A_{i+j\cdot2(L-1)+1} z^{-i} \right) \cdot \left(z^{-2(L-1)} \right)^j}{1 - z^{-1} \cdot \displaystyle\sum_{j=1}^{D} \left(B_{jL} z^{-\left(jL-2(L-1)\cdot(p_j-1)-1 \right)} \right) \cdot \left(z^{-2(L-1)} \right)^{(p_j-1)}} \qquad (2.14a)$$

where

$$B_{ncn} = \left\lfloor \frac{N + D(L-1) - 1}{2(L-1)} \right\rfloor \quad \& \quad B_{ncd} = \left\lfloor \frac{DL}{2(L-1)} \right\rfloor \qquad (2.14b)$$

and

$$p_j = \left\lfloor \frac{jL}{2(L-1)} \right\rfloor \qquad (2.14c)$$

It is obvious that a non-canonic structure requires fewer, though relatively high-speed, amplifiers due to the reduced number of ADB's and single accumulator. On the contrary, the canonic structure needs more, though slower, opamps, like in the FIR counterparts. More importantly, no mater non-recursive or recursive ADB structures are, both evolve from the DF polyphase prototype [2.11, 2.12, 2.13], thus will all succeed in the inherent immunity to the input lower-rate S/H shaping distortion.

4.2 SC Circuit Architectures

To generalize with simplicity, only SC circuitry for a recursive-ADB structure will be presented for the IIR interpolating filter, since the non-recursive ADB realization for the FIR function can be easily obtained only by removing the feedback recursive networks. Then, to illustrate the above, a lowpass interpolator for a video decoder will be used as an example, which converts a 3.6 MHz bandwidth composite analog video signal, from sampling at 10 MHz to 30 MHz. For standard CCIR 601 8-bit accuracy requirement, a 4th-order Elliptic filter with < 0.4 dB passband ripple and ≥ 40 dB attenuation is necessary. Its original (2.10) and multirate modified (2.11) transfer functions coefficients are listed in Table 2-1.

The corresponding canonic-form using R-ADB polyphase structures in a Complete-DFII (C-DFII) realization is shown in Figure 2-5, and after formulating its multirate transfer function through (2.13a) and (2.13b) the corresponding SC circuit is obtained and presented in Figure 2-6.

Table 2-1. Transfer function coefficients of 3-Fold SC LP IIR video interpolators:original (a_i and b_i) and multirate-transformed (A_i and B_i) for Elliptic and ER C-DFII structures

	Original			**Multirate Modified**	
(2.3)	**Elliptic** (*D*=4)	**ER** (*N*=9, *D*=2)	**(2.4)**	**Elliptic** (*D*=4)	**ER** (*N*=9, *D*=2)
a_0	0.0958	0.1006	A_0	0.0958	0.1006
a_1	0.0808	0.2146	A_1	0.2927	0.3181
a_2	0.1554	0.3616	A_2	0.5807	0.6198
a_3	0.0808	0.4385	A_3	0.8945	0.9613
a_4	0.0958	0.4084	A_4	1.1316	1.1926
a_5		0.2868	A_5	1.1983	1.2257
a_6		0.1355	A_6	1.0592	1.0613
a_7		0.0415	A_7	0.8073	0.7813
a_8		-0.0249	A_8	0.5250	0.4620
			A_9	0.2741	0.2199
			A_{10}	0.1140	0.0836
			A_{11}	0.0351	0.0020
			A_{12}	0.0065	-0.0117
b_1	2.2112	1.0285	B_3	-0.9688	-1.0256
b_2	-2.3148	-0.6850	B_6	-0.3683	-0.3214
b_3	1.1918		B_9	-0.0082	
b_4	-0.2695		B_{12}	-0.0175	

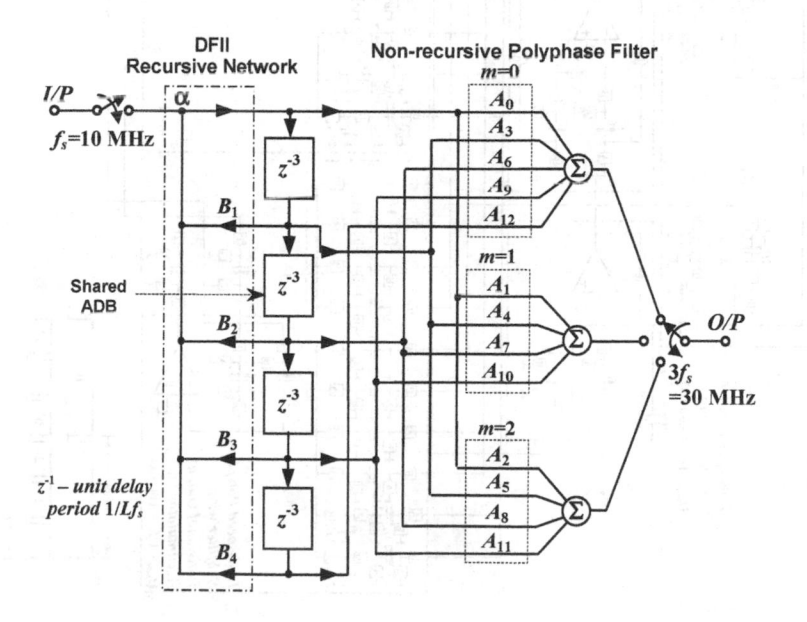

Figure 2-5. Canonic-form R-ADB/C-DFII polyphase structures for improved 3-fold SC IIR video interpolator

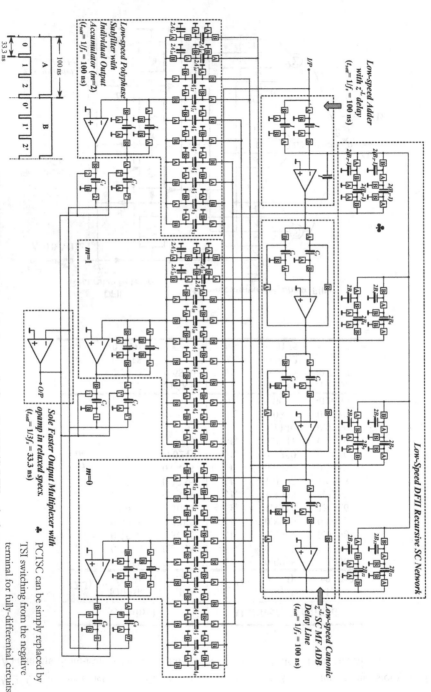

Figure 2-6. SC circuit schematic for canonic-form R-ADB/C-DFII polyphase structures

To further boost the speed capability of the filter, the double-sampling is efficiently employed due to the low-speed operation nature in canonic-form realization. The multirate denominator polynomials are obtained from the upper double-sampling DFII recursive feedback branches to the first specific adder stage that also implement simultaneously another functionality by embedding a z^{-3} delay. This adder/ADB together with the following SC ADB circuits form a low-speed serial delay line shared by both recursive and non-recursive networks. Especially, these L-unit ADB SC circuits exhibit a Mismatch-Free (MF) property for better reduction of the errors that will be accumulated along the delay line due to the finite gain and bandwidth, as well as the offset of the opamp when compared to the general charge-transferred delay circuit.

Considering one of the most efficient advantages of polyphase structures, namely the relaxed operation speed at the lower input sampling rate, the bottom half of the circuit contains $L=3$ low-speed DF polyphase filters by employing their corresponding individual slow accumulators, each being responsible for generating one of L output samples at lower input sampling rate. Thus, all the opamps in ADB's and accumulators have a very relaxed settling time requirement of full large input sampling period (100 ns), which is $L=3$ times longer than that of opamps, if a conventional bi-phase filter with double sampling was used. This also contributes to the reduction of the noise, charge injection and clock feedthrough errors in SC circuits.

The transfer function coefficients are implemented by either Toggle-Switched Inverter (TSI) and Parasitic-Compensated Toggle-Switched Capacitor (PCTSC) for positive and negative value respectively (PCTSC will be replaced by TSI switching from the negative terminal for fully-differential implementations which will be the dominant structure for state-of-the-art SC ICs). In addition, the recursive networks contribute not only to the common delay line but also to the non-recursive SC branches A_0, A_1 and A_2 for each polyphase subfilter at the same time, since input and recursive signals must originally be added together at node "α", as illustrated in Figure 2-5. In order to save this adder (one extra opamp) and to take advantage of both the existing output accumulator and of the opamp in the ADB, a coefficient-simplification procedure is proposed to each polyphase subfilter based on two sets of the same recursive networks – one that feeds back to the input of adder/ADB, and another that feeds forward to the output accumulator which can be efficiently combined together with existing non-recursive branches. In other words, no extra SC branches are needed, e.g., A_3 and A_6 in polyphase subfilter $m=0$ are simplified to $A_3' = B_3 \times A_0 + A_3$ and $A_6' = B_6 \times A_0 + A_6$ respectively, while A_4 in polyphase subfilter $m=1$ to $A_4' = B_3 \times A_1 + A_4$.

$L=3$ parallel double-sampling Toggle-Switched Capacitor (TSC) branches followed by an output unity-gain buffer can be simply used as a multiplexer (MUX) for switching the interpolated output from those three polyphase subfilters. Besides, there is another simple MF SC multiplexer which can employ the well-known fully-differential and bottom-plate sampling techniques to eliminate the signal-dependent charge-injection and clock feedthrough errors that are unavoidably existed in the aforementioned unity-gain buffer approach. Although it operates at higher output sampling rate (33.3 ns settling time – full output period), its feasibility is derived from the fact that specifications of the multiplexer opamp are much less stringent than those in ADB's or accumulators if operating at the same speed. This happens because the opamp always operates with a large feedback factor (> 0.5 when the sampling capacitor is greater than the input parasitics capacitance of opamp), thus reducing its bandwidth or transconductance requirements. Normally, the relatively smaller total equivalent capacitive loading compared with those formed by a set of coefficient capacitors (for opamps in ADB's) with also a large summing feedback capacitor (for opamps in accumulators), together with usually smaller output voltage step during two consecutive phases (due to the sampling rate increase nature), normally relax the opamp slew-rate and bandwidth requirements which are all directly proportional to the opamp power consumption. If it is necessary to drive a large capacitive load (like a pad of IC for testing purpose), then buffers with low output-impedance are normally required for better performance, because for higher power efficiency, opamp used in SC circuits are normally designed with high output impedance (also called transconductance opamp or Operational Transconductance Amplifier-OTA). Thus, especially in baseband lowpass systems, the power of this multiplexer opamp can be even smaller than those with wide settling time in ADB's (presented next). Moreover, the errors caused by finite-gain and offset of this MF multiplexer will introduce smaller deviation and mostly just a gain shift and a DC offset in the overall system response. Then, its elimination of charge-transfer reduces not only the mismatch error for each path but also the special glitches in the output signal caused by the opamp high output-impedance that normally appears in the beginning of the charge-transfer in transconductance-opamp-based SC circuits. Consequently, the canonic ADB structure is very attractive for high frequency operation.

By formulating the multirate-transformed transfer function of this 3-fold Elliptic interpolating filter from (2.14a), (2.14b) and (2.14c), the circuit can be also designed with non-canonic R-ADB polyphase structures in a C-DFII architecture, where the simplified structure and its corresponding SC circuit diagram are presented in Figure 2-7 and Figure 2-8, respectively.

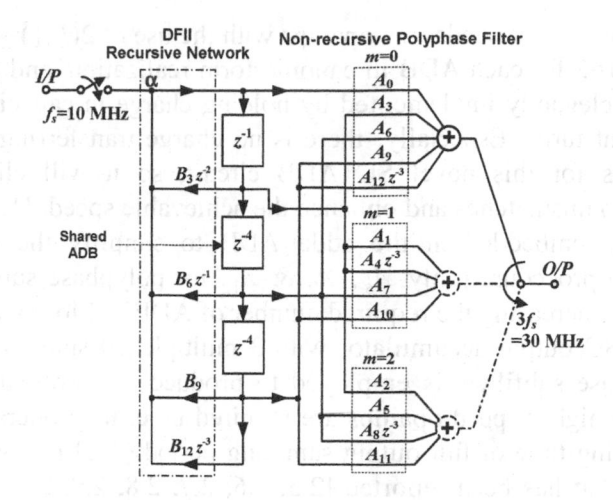

Figure 2-7. Non-canonic-form R-ADB/C-DFII polyphase structures for improved 3-fold SC IIR video interpolator

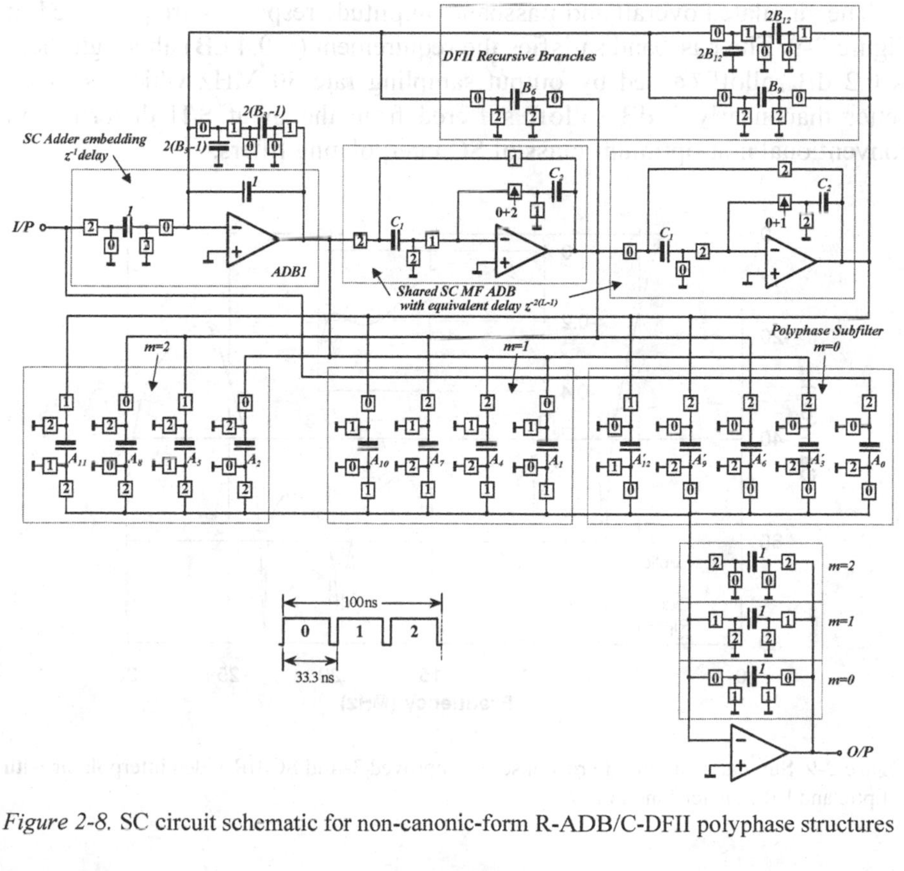

Figure 2-8. SC circuit schematic for non-canonic-form R-ADB/C-DFII polyphase structures

It is obvious that it needs less opamps with the use of $2(L-1)=4$ unit delay instead of $L-1=2$ for each ADB in canonic-form realization, and the extra 2-unit delay is elegantly implemented by holding charge in capacitors C_1 and C_2 in different turns. Especially, there is no charge transferring during the delay process for this novel SC ADB circuit, so it will eliminate the capacitor ratio mismatches and enhance the achievable speed. Here, only the unity-delay is embedded in the adder/ADB to simplify the coefficient-simplification procedure (only A_6, A_9 & A_{12} for polyphase subfilter $m=0$) while without increasing the required number of ADB's. Moreover, only one time-shared SC output accumulator with 3 multiplexed summing branches for 3 polyphase subfilters is employed to produce L interpolated outputs. Although the higher-speed opamps are required here, they operate with the required settling time of full output sampling period (33.3 ns) which is still wider than what has been reported [2.5, 2.6, 2.7, 2.8, 2.9, 2.10]. The total number of opamps will be saved to 4 only and the sensitivity performance will also be improved when compared to those of the canonic realization, thus it is more suitable for lower speed applications.

The simulated overall and passband amplitude responses are presented in Figure 2-9. The passband satisfies the requirement (< 0.4 dB) although there is 0.2 dB rolloff caused by output sampling rate 30 MHz which is much better than nearly 2 dB rolloff suffered from the input S/H distortion in conventional non-optimum-class of SC interpolating filters.

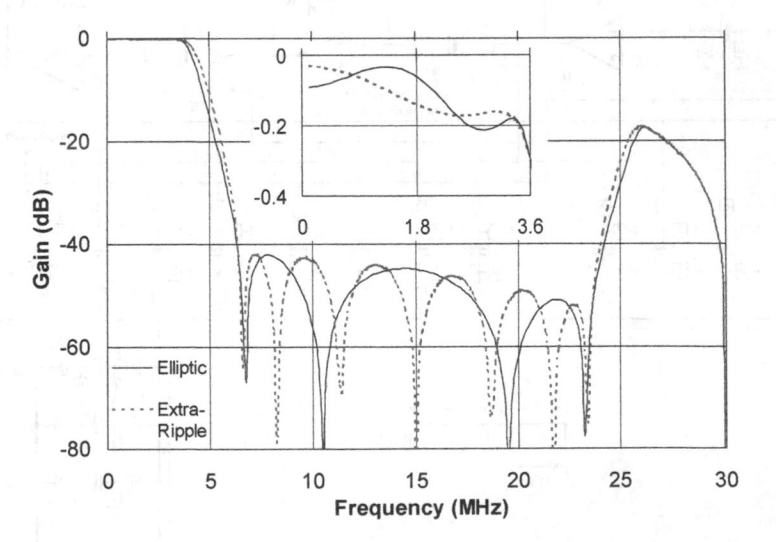

Figure 2-9. Simulated amplitude response for improved 3-fold SC IIR video interpolator with Elliptic and ER transfer function

5. LOW-SENSITIVITY MULTIRATE IIR STRUCTURES

5.1 Mixed Cascade/Parallel Form

Although high-order IIR interpolators can be implemented directly in a single stage by employing the above R-ADB/C-DFII polyphase structures, cascade or parallel form structures are usually preferable for their lower sensitivity to coefficient deviation. Therefore, for interpolation with relatively smaller or prime L factors but higher IIR filter order, Parallel Form (P-DFII) structures can be simply achieved by expressing the rational transfer function in a partial fraction expansion and implementing it by the 1st- and 2nd-order building blocks in parallel. Thus, the corresponding modified multirate transfer function can be expressed as

$$\hat{H}(z) = \sum_{i=1}^{S} \frac{\hat{N}_{P\_i}(z)}{\hat{D}_i(z)} \tag{2.15}$$

where S is the number of the stages and each stage can be realized by the above DFII R-ADB polyphase structures.

Nevertheless, the cascade form has normally better sensitivity performance than parallel form due to the independence of the errors in each section caused by their poles and zeros deviation, while the sensitivity performance of parallel form highly depends on the output adder. However, the pure cascade form is actually a multistage implementation of interpolation, that is only suitable for large or nonprime alteration factor L due to its inherent nonidentity of input and output sampling rate. Therefore, here another alternative is proposed: Mixed Cascade/Parallel (MCP-DFII) structure, which is a combination of a cascade of low-order recursive DFII parts and a multi-feed-out parallel non-recursive polyphase subfilter structure (designated as internally-cascaded [2.22]) and is especially suitable for sampling rate conversion. Since the cascade of recursive parts leads to a considerably large reduction in the dependency between coefficient sensitivity and output adder, it improves significantly the overall circuit sensitivity performance. In this case, the modified transfer function can be mathematically decomposed into the following form

$$\hat{H}(z) = \sum_{i=1}^{S} \left(T_i(z) \cdot \frac{\hat{N}_{MCP\_i}(z)}{\prod_{j=1}^{i} \hat{D}_j(z)} \right) \tag{2.16}$$

where the $T_i(z)$ is the accumulated delay factor introduced by the cascade of recursive parts and $T_1(z) = 1$. An optimized choice of this delay factor will render a better performance, and the idea is actually to lower the quality factor of each cascaded stage.

This structure can be further explained by considering the application to the same 3-fold 4th-order video interpolator. The coefficients of the modified multirate transfer function for P-DFII and MCP-DFII ($T_2(z) = z^{-6}$) realizations are all tabulated in Table 2-2 and their corresponding R-ADB polyphase structures are shown in Figure 2-10(a) and (b) for P-DFII and MCP-DFII, respectively. As will be illustrated later, MCP-DFII structure offers a much better performance than C-DFII and P-DFII especially for high order functions. Hence, we only present in Figure 2-11 the SC implementation of MCP-DFII structure in non-canonic form for simplicity and comparison with the previous circuits, although canonic-form is also equivalently applicable, as well as the P-DFII can also be derived similarly. The output accumulators of the polyphase filters in these two 2nd-order sections are efficiently shared for reduced number of opamps. Furthermore, both P-DFII and MCP-DFII always offer an extra superiority in reducing the capacitor spread, e.g. Maximum coefficient spread for C-DFII, P-DFII and MCP-DFII are 209, 67 and 58, respectively, in this example. The simulated results are the same as C-DFII as shown in Figure 2-9.

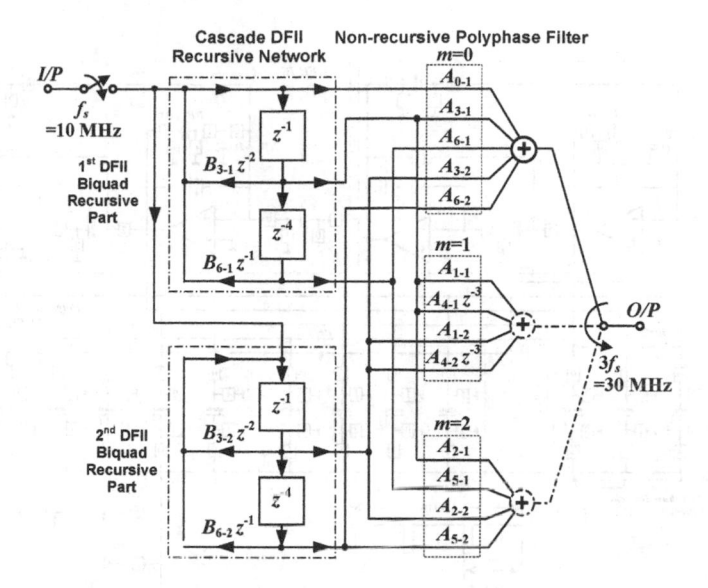

Figure 2-10(a). R-ADB/P-DFII for Improved 3-fold SC IIR video interpolator

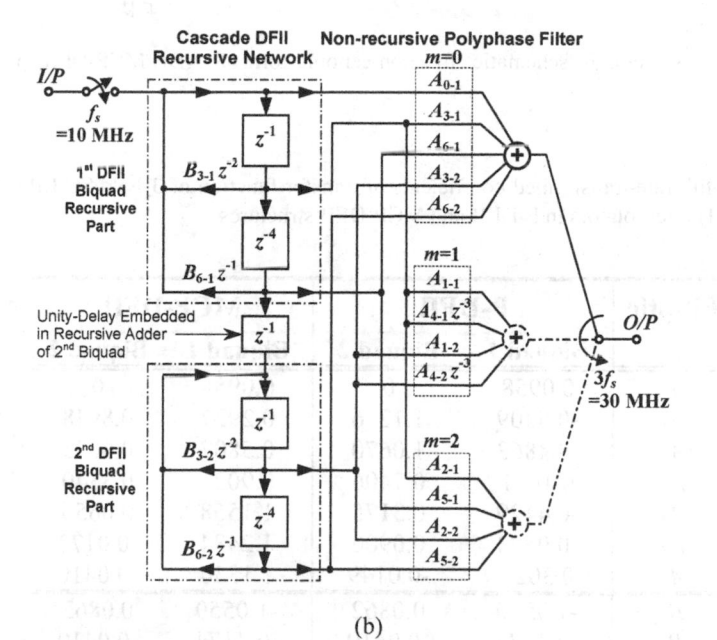

(b)

Figure 2-10(b). R-ADB/MCP-DFII for Improved 3-fold SC IIR video interpolator

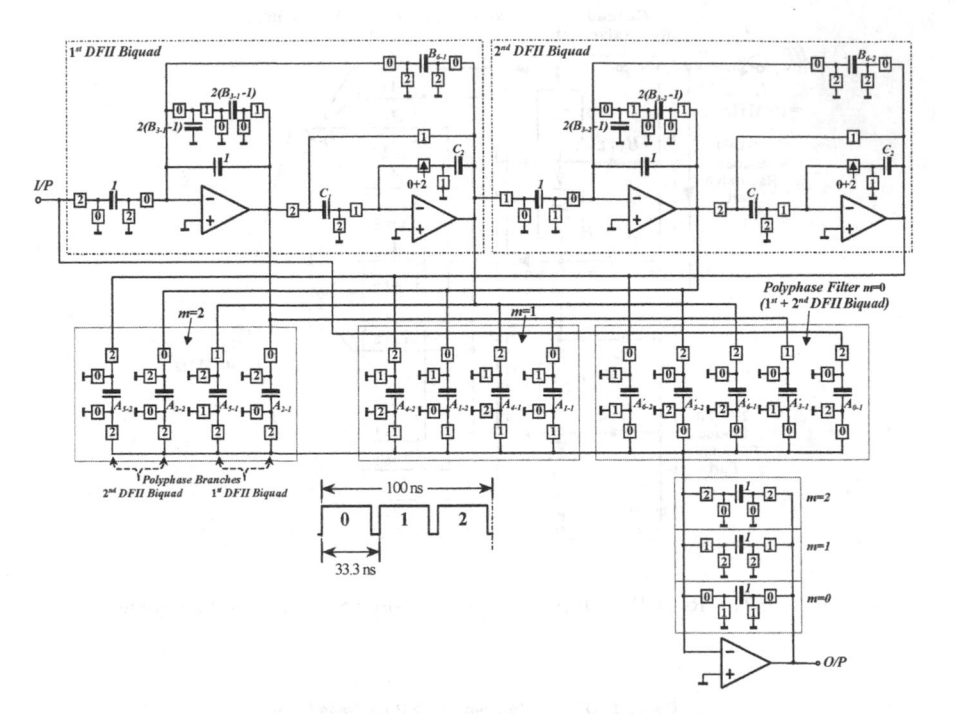

Figure 2-11. SC circuit schematic for non-canonic-form R-ADB/MCP-DFII polyphase structures

Table 2-2. Multirate-transformed coefficients of transfer function of 3-Fold SC LP IIR video Elliptic (D=4) interpolators in P-DFII and MCP-DFII structures

Elliptic	P-DFII		MCP-DFII	
	Biquad 1	Biquad 2	Biquad 1	Biquad 2
A_0	0.0958	0	0.0958	0
A_1	-0.8309	1.1236	0.2927	0.8948
A_2	-0.4863	1.0670	0.5807	0.6083
A_3	0.1621	0.7406	0.9027	0.3340
A_4	-0.4429	0.3175	1.1568	0.0656
A_5	-0.0593	0.0900	1.2484	-0.0172
A_6	0.3027	-0.0149	1.1330	-0.0410
B_3	-1.0550	0.0862	-1.0550	0.0862
B_6	-0.4174	-0.0419	-0.4174	-0.0419

5.2 Extra-Ripple IIR Form

Another alternative technique for IIR interpolation uses the Extra-Ripple (ER) type IIR transfer function obtained by the improved Martinez/Parks algorithm [2.23] for achieving better sensitivity in passband due to its advantage of smaller denominator order, by optimum positioning of the poles and zeros [2.23, 2.24]. For the same specifications of the above video interpolator, the original ER IIR transfer function is obtained with only lower 2nd-order denominator but at the price of a higher 8th-order numerator (N=9, D=2). However, its multirate-transformed transfer function, with coefficients shown in Table 2-1 together with the original, has a denominator order of 6, but, more importantly, exactly the same order of 12 in the numerator when compared with that of the 4th-order IIR Elliptic, as shown also in Table 2-1. This means that no penalty is present for increasing the number of zeros and that shows its additional superiority for the use in multirate circuits. It has an identical implementation in R-ADB/C-DFII structure with either canonic or non-canonic form, as in Figure 2-6 and Figure 2-8, but with 2 less recursive branches. If higher denominator order is required, both P-DFII and MCP-DFII realizations can also be preferably employed. The simulated results for their corresponding SC circuits in both canonic and non-canonic forms are the same and illustrated in the dashed curve of Figure 2-9.

6. SUMMARY

This chapter presents first the rigorous mathematical analysis on conventional sampled-data analog interpolation whose response is distorted by undesired input lower-rate S/H shaping effect. A new ideal improved analog interpolation model has then been presented to entirely eliminate such distortion over the whole frequency axis. Both traditional Bi-phase SC structures and multirate polyphase structures have been described in order to achieve such improved analog interpolation. Especially, the multirate polyphase structure has been proven to be an efficient and effective realization for optimum-class analog interpolation with respect to the competent power and silicon consumption. Different ADB polyphase-based structures with their corresponding SC architectures have then been investigated thoroughly for practical higher-order filtering functions: FIR non-recursive ADB and IIR recursive ADB in their canonic and non-canonic realizations with L low-speed accumulator and single time-shared accumulator schemes, respectively. Detailed practical IC design considerations and different structures' pros and cons will be further studied and analyzed next.

REFERENCES

[2.1] C.W.Solomon, L.Ozcolak, G.Sellani, W.E.Brisco, "CMOS analog front-end for conversational video phone modem," in *Proc. IEEE Custom Integrated Circuits Conference (CICC)*, pp.7.4/1-5, 1989.

[2.2] D.Senderowicz, G.Nicollini, P.Confalonieri, C.Crippa, C.Dallavalle, "PCM Telephony: Reduced architecture for a D/A converter and filter combination," *IEEE J. Solid-State Circuits*, vol. 25, pp.987–995, Aug.1990.

[2.3] B.Baggini, L.Coppero, G.Gazzoli, L.Sforzini, F.Maloberti, G.Palmisano, "Integrated digital modulator and analog front-end for GSM digital cellular mobile radio system," *Proc. IEEE Custom Integrated Circuits Conference (CICC)*, pp.7.6/1 -7.6/4, 1991.

[2.4] C.S.Wong, "A 3-V GSM baseband transmitter," *IEEE J. Solid-State Circuits*, vol.34, No.5, pp.725-730, May 1999.

[2.5] J.E.Franca, "Non-recursive polyphase Switched-Capacitor decimators and interpolators," *IEEE Trans. Circuits and Systems*, Vol. CAS-32, pp. 877-887, Sep.1985.

[2.6] R.P.Martins, J.E.Franca, "Infinite impulse response Switched-Capacitor interpolators with optimum implementation", in *Proc. IEEE International Symposium on Circuits and Systems (ISCAS)*, pp.2193-2196, May 1990.

[2.7] R.P.Martins, J.E.Franca, "Novel second-order Switched-Capacitor interpolator", *IEE Electronics Letters*, Vol.28 No.2, pp.348-350, Feb.1992.

[2.8] C.-Y.Wu, S.Y.Huang, T.-C.Yu, Y.-Y.Shieu, "Non-recursive Switched-Capacitor decimator and interpolator circuits," in *Proc. IEEE International Symposium on Circuits and Systems (ISCAS)*, pp.1215-1218, 1992.

[2.9] K.Kato, T.Kikui, Y.Hirata, T.Matsumoto, T.Takebe, "SC FIR interpolation filters using parallel cyclic networks," in *Proc. IEEE International Symposium on Circuits and Systems (ISCAS)*, pp.723-726, 1994.

[2.10] P.J.Santos, J.E.Franca, "Switched-capacitor interpolator for direct-digital frequency synthesizers," in *Proc. IEEE International Symposium on Circuits and Systems (ISCAS)*, Vol.2, pp.228 -231, 1998.

[2.11] U Seng Pan, R.P.Martins, J.E.Franca, "Switched-Capacitor interpolators without the input sample-and-hold effect," *IEE Electronics Letters*, Vol.32, No.10, pp.879-881, May 1996.

[2.12] Seng-Pan U, *Impulse Sampled Switched-Capacitor Sampling Rate Converters*, Master Thesis, University of Macau, Macao SAR, China, 1997.

[2.13] Seng-Pan U, R.P.Martins, J.E.Franca, "Improved Switched-Capacitor interpolators with reduced sample-and-hold effects," *IEEE Trans. Circuits and Systems – II: Analog and Digital Signal Processing*, Vol.47, No.8, pp.665-684, Aug. 2000.

[2.14] R.E.Crochiere, L.R.Rabiner, *Multirate Digital Signal Processing*, Prentice-Hall, Inc., NJ, 1983.

[2.15] S.K.Mitra, J.F.Kaiser, *Handbook for Digital Signal Processing*, John Wiley & Sons, Inc., 1993.

[2.16] J.E.Franca, A.Petraglia, S.K.Mitra, "Multirate analog-digital systems for signal processing and conversion," *Proc. of The IEEE*, Vol.85, No.2, pp.242-262, Feb.1997.

[2.17] J.E.Franca, S.Santos, "FIR Switched-Capacitor Decimators with Active-Delayed Block Polyphase Structures," *IEEE Trans. Circuits and Systems*, Vol. CAS-35, pp.1033-1037, Aug. 1988.

[2.18] Seng-Pan U, R.P.Martins, J.E.Franca, "Impulse sampled FIR interpolation with SC Active-Delayed Block polyphase structures," *IEE Electronics Letters*, Vol.34, No.5, pp.443-444, Mar.1998.

[2.19] J.E.Franca, R.P.Martins, "IIR Switched-Capacitor decimator building blocks with optimum implementation," *IEEE Trans. Circuits and Systems*, Vol. CAS-37, No.1, pp.81-90, Jan. 1990.

[2.20] R.P.Martins, J.E.Franca, F.Maloberti, "An optimum CMOS Switched-Capacitor antialiasing decimating filter," *IEEE J. Solid-State Circuits*, Vol.28 No.9, pp.962-970, Sep. 1993.

[2.21] U Seng Pan, R.P.Martins, J.E.Franca, "New impulse sampled IIR Switched-Capacitor interpolators," in *Proc. IEEE International Conference on Electronics, Circuits and Systems (ICECS)*, pp.203-206, Oct.1996.

[2.22] R.P.Martins, J.E.Franca, "Design of cascade Switched-Capacitor IIR decimating filters," *IEEE Trans. on Circuits and Systems–I*, Vol.42, No.7, pp.367-376, Jul.1995.

[2.23] L.B.Jackson, "An Improved Martinez/Parks Algorithm for IIR Design with Unequal Numbers of Poles and Zeros," *IEEE Trans. on Circuits and Systems*, Vol.42, No.5, pp.1234-1238, May 1994.

[2.24] A.Petraglia, J.S.Pereira, "Switched-Capacitor decimation filters with direct form polyphase structure having very small sensitivity characteristics," in *Proc. IEEE International Symposium on Circuits and Systems (ISCAS)*, Vol.II, pp.73-76, May.1999

Chapter 3
PRACTICAL MULTIRATE SC CIRCUIT DESIGN CONSIDERATIONS

1. INTRODUCTION

To implement successfully in silicon the proposed SC architectures presented before and for achieving optimum-class interpolation filtering comprehensive practical design considerations are investigated in this chapter by focusing on the power efficiency of canonic and non-canonic SC structures together with the associated imperfections resulting from capacitance ratio inaccuracies, finite-gain and bandwidth and input-referred DC offset effects of the opamps as well as the clock random jitter and timing-skew effects. Finally, a simple noise analysis methodology for the polyphase-based interpolating filters will be also presented.

2. POWER CONSUMPTION ANALYSIS

In order to estimate the approximate analog power dissipated in the proposed SC interpolating filters for both canonic and non-canonic forms, we will use the single-stage telescopic transconductance opamp architecture, which is often used for high-speed applications. The equivalent continuous-time model of an SC circuit during the charge transfer phase (e.g. in either phase A or B of Figure 2-3), as shown in Figure 3-1, with a simplified single-pole transconductance opamp model, is a good approximation of a single-stage transconductance opamp with the phase margin > 60°. C_I and C_L are, respectively, the total capacitance of input and output SC branches connected to the opamp in this phase. Assuming that 1/5 of the phase duration is allocated for slewing while the remaining 4/5 for linear settling

($t_{sett} = t_{slew} + t_{lin}$), for the worst-case estimation, the opamp must be capable to drive the equivalent total capacitive loading C_{Ltot} to a certain output voltage step V_{ostep} during the slewing time interval t_{slew}, and also to be settled within 0.1 % accuracy during the subsequent time interval t_{lin} (the closed-loop time constant is approximately $t_{lin} / 7$). Thus, the required tail bias current I_{SS} of the opamp can be simply estimated as

$$I_{SS} = \max\left(I_{SS\_SR}, I_{SS\_lin}\right) \tag{3.1}$$

where I_{SS_SR} and I_{SS_lin} are the bias current required in slewing and linear settling time intervals, respectively, given by

$$I_{SS\_SR} = SR \cdot C_{Ltot} = \frac{\left|V_{ostep}\right|}{t_{slew}} C_{Ltot} \tag{3.2}$$

$$I_{SS\_lin} = g_m \cdot V_{eff} = \frac{V_{eff} \cdot C_{Ltot}}{\beta \cdot (t_{lin}/7)} \tag{3.3}$$

where SR and g_m are the required slew rate and transconductance, V_{eff} is the effective or overdrive voltage for differential-pair MOS transistors, and the equivalent total capacitive loading

$$C_{Ltot} = C_L + C_{PO} + \beta \cdot (C_I + C_{PI}) \tag{3.4}$$

with the feedback factor

$$\beta = \frac{C_F}{C_I + C_{PI} + C_F} \tag{3.5}$$

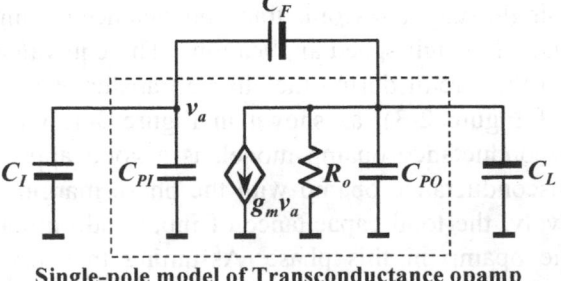

Single-pole model of Transconductance opamp

Figure 3-1. Equivalent continuous-time model of SC circuit during charge-transfer phase

Therefore, the expected static power of the opamp is obtained by multiplying the supply voltage and the I_{SS}, i.e. $P_{OTA} = V_{DD}I_{SS}$. Normally, an optimum solution of the bias current should be investigated in the real design according to the required specifications in terms of gain, speed, power, dynamic range and noise. However, the above estimation is still very useful for an initial stage of the design. And, as it will be presented next, since SC interpolating filter is typically not applied in the oversampling case, the required tail bias current will be mostly dominated by the I_{SS_SR}.

According to the above expressions, the approximate analog power of the 3-fold IIR LP interpolator with ER transfer function in canonic and non-canonic forms introduced in Chapter 2 / Session 5.2 is presented in Table 3-1 (V_{eff} is typically assumed to be 200 mV). For a more realistic approach, each circuit here uses only one fast and one slow opamp instead of several different speed opamps. Although canonic form has the double number of opamps when compared with the non-canonic form, the power is still about only 58 % of the latter due to the very low speed operation of these opamps. Thus, this proves again that <u>the canonic form is very attractive for high-frequency applications</u> not only from the perspective of <u>power efficiency</u> but mainly from a much more <u>relaxed design in lower speed opamp</u>. Especially, the higher-speed opamp in the output multiplexer in canonic structure consumes not much power or even less when compared to those opamps with L times enlarged t_{sett} in ADB's and accumulators. Also, it is obvious that the opamp in the multiplexer always needs less power and also smaller g_m than those in ADB's and accumulators with the same t_{sett} in non-canonic form.

Table 3-1. Power comparison for 3-Fold SC LP IIR with ER transfer function

	IIR Canonic				IIR Non-Canonic		
OTA	*R. Adder* ×1	*MF ADB* ×3	*Accu.* ×3	*O/P Mul.* ×1	*R. Adder* ×1	*MF ADB* ×2	*Accu.* ×1
t_{sett}	100 ns	100 ns	100 ns	33.33 ns	33.33 ns	33.33 ns	33.33 ns
V_{ostep}	1 V	1 V	1 V	0.6 V	1 V	1 V	0.6 V
SR	50 V/μs	50 V/μs	50 V/μs	90 V/μs	150 V/μs	150 V/μs	90 V/μs
g_m	2.92 mS	0.65 mS	0.6 mS	0.65 mS	4.94 mS	2.7 mS	3.38 mS
I_{SS}	0.58 mA	0.22 mA	0.13 mA	0.13 mA	0.99 mA	0.73 mA	0.68 mA
No. of Use	×2			×6	×2		×2
Total Power	5.8 mW (3 V Supply)				10 mW (3 V Supply)		

Note: The highest g_m and I_{SS} are presented for the opamps in ADB's and accumulators.

3.   CAPACITOR-RATIO SENSITIVITY ANALYSIS

The sum-sensitivity [3.1] of the response with respect to all capacitors is performed to compare different designs with deviations of multiple circuit parameters.

3.1   FIR Structure

For a more general case, we consider an example of an 18-tap LP 4-fold interpolator, whose amplitude sum-sensitivity for the ADB polyphase structures in both canonic and non-canonic forms are performed as shown in Figure 3-2(a). Canonic realization has relatively worse overall sensitivity when compared with non-canonic mainly due to the double-sampling nature. For its only-zero property, FIR structure presents a good sensitivity in the passband but relatively poor sensitivity in the stopband. For a 0.3 % capacitor ratio error which can be achieved in current technologies, maximum passband and stopband deviations are roughly 0.025 & 2.7 dB and 0.018 & 2 dB for canonic and non-canonic form, respectively, which still have a satisfactory more than -40 dB attenuation (7 to 8-bit accuracy) in the stopband. Since the coefficients are implemented with direct capacitor ratio, the stopband is also predictable to have the mean about -43 dB from the estimation by the sum of original stopband ripple with the mean value of the expected magnitude deviation $\bar{h}_k \sigma_e \cdot \left(\sqrt{\pi N}/2 \right)$ ($\bar{h}_k$ – arithmetic mean value of all coefficients; σ_e – standard deviation of ratio error) for an N-Tap FIR filter obtained by the Rayleigh distribution [3.2, 3.3, 3.4]. In addition, we propose here also a further estimation to the worst-case stopband, i.e. about -41 dB, by using $h_{k\max}$ (passband normalized to 1) instead of $\bar{h}_k$, to approximate the worst-case magnitude deviation, i.e.

$$\mu_{|\Delta G|}(\omega)_{wc} = h_{k\max} \sigma_e \cdot \left(\sqrt{\pi N}/2 \right) \tag{3.6}$$

This has been verified with a good agreement by comparing it to the Monte-Carlo simulation shown in Figure 3-2(b) with respect to all coefficients which are independent zero-mean Gaussian random variables with $\sigma_e = 0.3$ %. Thus, from the above prediction expressions and also the Monte-Carlo simulation, an SC 50-tap 4-fold FIR interpolator with theoretical -45.5 dB stopband (for regular LP L-fold interpolation, $h_{k\max} \approx 1/L$) can achieve the worst-case stopband about 7 to 8-bit accuracy with $\sigma_e = 0.3$ % (without counting other non-ideal effects in SC realization). It is also expected that it will be quite difficult to achieve higher than 8-bit accuracy for high order SC FIR filter without specific improvement techniques.

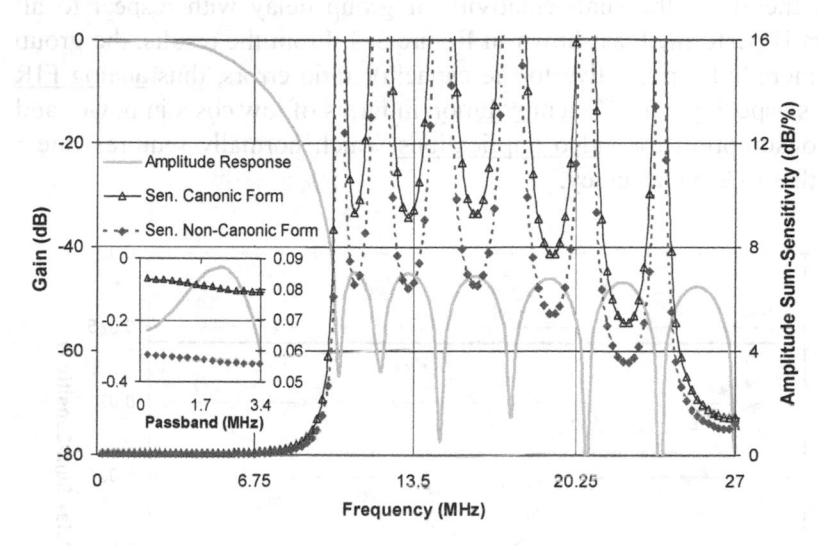

(a)

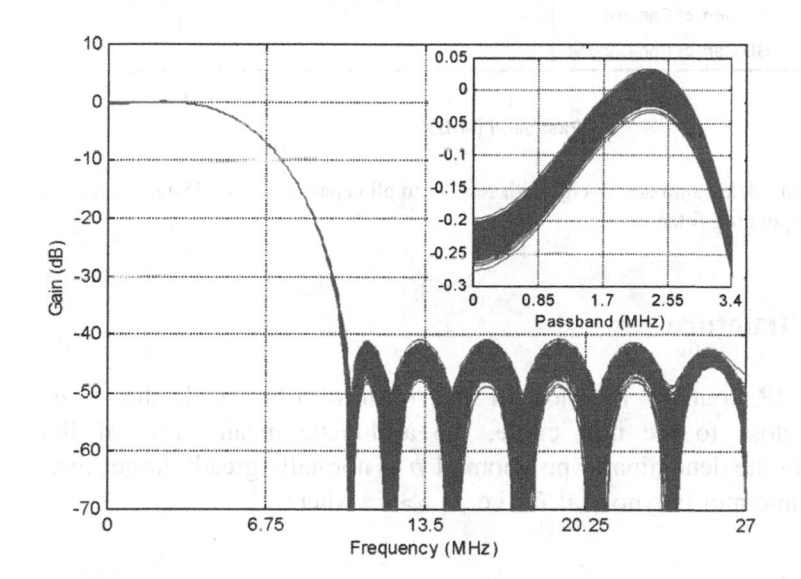

(b)

Figure 3-2. (a) Amplitude sum-sensitivity (b) Monte-Carlo simulations with respect to all capacitors of an 18-tap improved SC FIR LP interpolating filter

One attractive advantage of FIR implementation is its linear phase property, therefore, the sum-sensitivity of group delay with respect to all capacitors is performed, as shown in Figure 3-3. From the results, the group delay is incredibly insensitive to the capacitor ratio errors, thus <u>analog FIR filtering</u> is especially an <u>efficient solution</u> in terms of low costs in power and silicon consumption for <u>video applications</u> which normally requires linear phase with 7 to 8-bit accuracy.

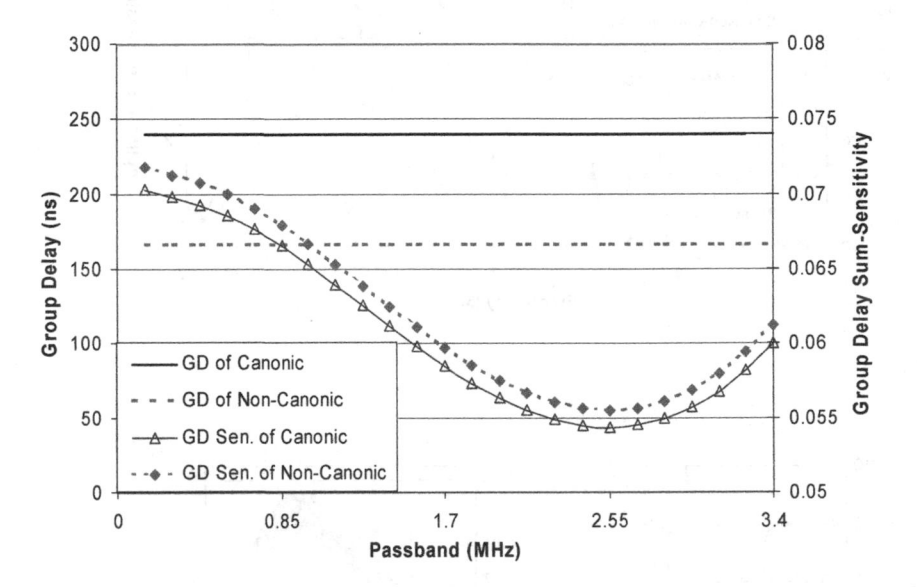

Figure 3-3. Group-delay sum-sensitivity with respect to all capacitors of an 18-tap improved SC FIR LP interpolating filter

3.2 IIR Structure

For the IIR transfer function in the form of (2.4) with the poles sufficiently close to the unit circle, the arithmetic mean value of the coefficients in the denominator polynomial $\overline{b}$ is normally greatly larger than that of the numerator polynomial $\overline{a}$, i.e. $\overline{b} \gg \overline{a}$, where

$$\overline{a} = \frac{1}{N} \sum_{k=0}^{N-1} |a_k| \qquad\qquad (3.7a)$$

and $\bar{b} = \dfrac{1}{D}\displaystyle\sum_{k=0}^{D}|b_k|$ (3.7b)

As a result, the expected value of the magnitude of the deviation in the frequency response for frequencies in the passband and stopband can be approximately obtained respectively by [3.3],

$$\mu_{|\Delta H|}(\omega)_{passband} = \frac{\bar{b}\,\sigma_e\sqrt{\pi D}}{2|B(\omega)|}$$ (3.8a)

$$\mu_{|\Delta H|}(\omega)_{stopband} = \frac{\bar{a}\,\sigma_e\sqrt{\pi N}}{2|B(\omega)|}$$ (3.8b)

where σ_e is the standard deviation of the ratio error.

For further verification, the simulated sum-sensitivity of the 4th-order IIR video interpolator presented before with non-canonic in C-DFII, P-DFII and MCP-DFII, as well as C-DFII with ER transfer function (C-DFII/ER), respectively, are presented in Figure 3-4(a). As expected, C-DFII/ER obtains the best sensitivity in the passband due to its less number of poles, and its stopband has a similar level when compared to all other realizations because there are no extra zeros in the multirate form. The MCP-DFII, which remains superior to the cascade structure, is more advanced in the overall response than the P-DFII whose performance depends on the output adder. However, these two are both worse than C-DFII in the passband since poles are not tightly clustered, due to the relatively lower order and larger transition band, thus the low sensitivity advantage of cascade or parallel structures is not explicit. This can be observed in a higher 6th-order IIR interpolating filter whose simulated sensitivities are shown in Figure 3-4(b). Both P-DFII and MCP-DFII are much less sensitive than C-DFII in the passband, stopband and also pole-zero cancellation, and the MCP-DFII achieves the best performance, as expected.

Besides, an ER IIR interpolator ($N=9$, $D=4$) for the same specifications is again much less sensitive when compared with the 6th-order interpolator both in C-DFII realization, while its MCP-DFII structure possesses a performance similar to the 6th-order implemented also with MCP-DFII. This results from the fact that it still requires 4 poles to maintain the flatness in this relatively wide passband. However, comparing that with the general IIR transfer function ($N-1=D$), ER form is still a good alternative especially for multirate filtering due to its reduced passband sensitivity (less number of poles) without increasing sensitivity in the stopband in most cases (similar number of zeros in multirate form), and its superiority will be very apparent for narrow passband, i.e. $D=2$. The small sensitivity overshoots nearly half of higher output sampling rate are caused by the incomplete multirate pole-zero cancellation.

In addition, for the same reason as in its FIR counterpart, canonic structures are more sensitive than the non-canonic. Besides, the delay factor $T_i(z)$ in MCP-DFII is important as aforementioned, e.g. the sensitivity has an 18 % increase if $T_2(z) = z^{-3}$ (use 1st-delay block output for 2nd DFII biquad input) like in this example.

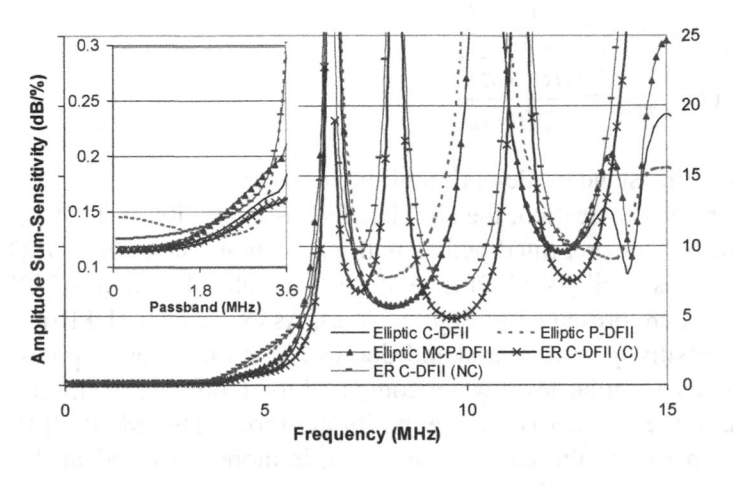

(a)

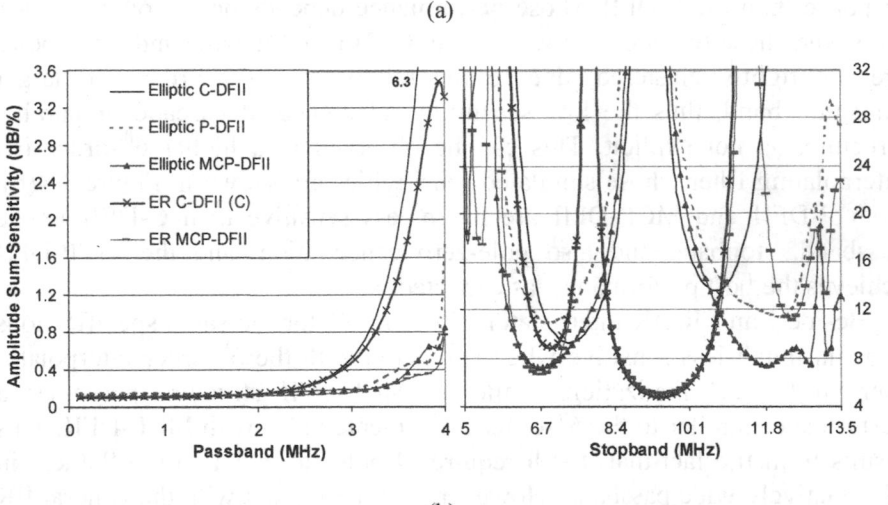

(b)

Figure 3-4. Amplitude sum-sensitivity with respect to all capacitors for improved 3-fold SC IIR video interpolating filter with different architectures and with (a) 4th-Order Elliptic & ER (N=9, D=2) and (b) 6th-Order Elliptic & ER (N=9, D=4) transfer functions

4. FINITE GAIN & BANDWIDTH EFFECTS

The practical finite gain and bandwidth of opamps will mainly lead to a system response deviation. As an example, it will be considered here the 3-fold interpolator with ER IIR transfer function due to its FIR-like multi-notch stopband. The simulated results using the opamp model from Figure 3-1, with a gain of 3000 and a nominal g_m (gm_nm) in Table 3-1, are presented in Figure 3-5(a) and (b) for, respectively, passband and stopband with either keeping the nominal g_m (same speed) but reducing the gain to 500 or keeping a low gain of 500 but with extra 40 % reduction in nominal g_m. Results show that passband deviation imposed by finite gain of opamps is less sensitive than that caused by bandwidth of opamps as the former leads to an almost net gain shift while the latter to a relatively larger rolloff in the passband. Although these errors lead to the movement of zeros from the unit circle, affecting the stopband and also the cancellation of poles and zeros, as shown in Figure 3-5(b) and especially around half of the output sampling rate, 40 dB attenuation is still achieved. The situation of the canonic structure is also worse than the non-canonic, but the low-speed & low-power requirements of the former allow to have a free headroom in design and also a decreased sensitivity to process variation. Moreover, the errors due to the finite gain effect will be further analyzed rigorously in the next chapter.

5. INPUT-REFERRED OFFSET EFFECTS

The input-referred offset errors will result in a reduced Signal-to-Noise-Ratio (SNR) due to the undesired fixed pattern noise placed at lower input sampling rate and its multiples particularly due to the low-speed operation nature of the optimum-class multirate interpolation. For ADB polyphase-based structures, those offset errors are mainly sourced from opamp DC offset propagation and accumulation along the serial ADB delay line in addition to the opamp DC offset mismatches among parallel polyphase subfilters especially for canonic realization, as well as the charge injection & clock-feedthrough effects due to the non-ideal analog switches.

Due to the parallel nature of polyphase-subfilter structures and considering that the overall offset error of each parallel subfilter is O_m ($m=0,1,\ldots,M$-1, where M is the parallel path number normally equal to the interpolation factor L for standard configuration of parallel polyphase structure) the discrete-time output signal spectrum with a sine wave input signal $A\sin(\omega_{in}t)$ can be expressed as [3.4, 3.5, 3.6, 3.7, 3.8, 3.9]

$$Y_d(\omega) = Y_s(\omega) + Y_{os}(\omega) \tag{3.9}$$

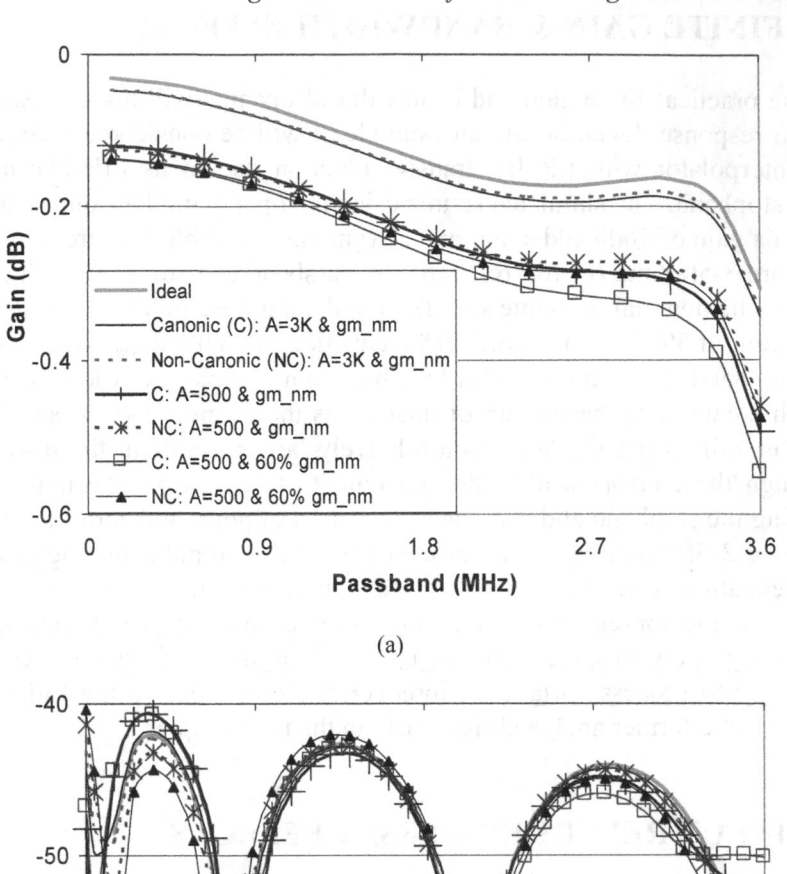

(a)

(b)

Figure 3-5. Opamp finite gain & bandwidth effects for improved 3-fold SC IIR interpolator
with ER (N=9, D=2) transfer function (a) Passband (b) Stopband

and the final output continuous-time S/H signal spectrum will be the output-rate sin$x$/x-shaped version of $Y_d(\omega)$. From (3.9), the first term corresponds to the input signal while the second term to the distortion caused by subfilter offsets, and they can be expressed as

$$Y_s(\omega) = \frac{\pi A j}{T_o} \sum_{k=-\infty}^{\infty} \left(\delta(\omega + \omega_{in} - 2\pi k) - \delta(\omega - \omega_{in} - 2\pi k) \right) \qquad (3.10)$$

and

$$Y_{os}(\omega) = \frac{2\pi}{T_o} \sum_{k=-\infty}^{\infty} A_k \delta\left(\omega - \frac{2\pi k}{MT_o} \right) \qquad (3.11a)$$

$$A_k = \frac{1}{M} \sum_{m=0}^{M-1} O_m e^{-jkm\frac{2\pi}{M}} \qquad (3.11b)$$

where the T_o is the output sampling period, i.e. $1/Lf_s$. From (3.11), it is obvious that the resulting distortion, namely, fixed pattern noise, is signal independent and located at the lower input sampling rate and its multiples, i.e. mLf_s / M. Assuming that the subfilter offsets O_m are both independent Gaussian random variables with zero mean and a standard deviation of σ_m (m=0,1,...,M-1), and for simplicity, if $\sigma_m = \sigma_{os}$, then the expected value of magnitude of these noise components can be obtained by

$$E[|A_k|] = \frac{\sigma_{os}}{2} \sqrt{\frac{\pi}{M}} \qquad (3.12)$$

with its standard deviation given by $\sqrt{(4-\pi)/M} \cdot \sigma_{os}/2$. Moreover, by using the Parseval's relation we can also derive the expected total output pattern-noise power as

$$P_{os} = E\left[\frac{1}{L} \sum_{m=0}^{L-1} |A_k|^2 \right] = E\left[\frac{1}{L} \sum_{m=0}^{L-1} O_m^2 \right] = \frac{1}{L} \sum_{m=0}^{L-1} \sigma_m^2 \qquad (3.13)$$

or according to the assumption $\sigma_m = \sigma_{os}$, it can simply be approximated by

$$P_{os} = \sigma_{os}^2 \qquad (3.14)$$

with its standard deviation of $\sqrt{2/L} \cdot \sigma_{os}^2$. Although the magnitude of the pattern-noise tones is dependent of the path number M, the mean of the total pattern-noise power is independent of that, but the most important feature is related with the fact that both of them are completely independent of the input signal levels.

Considering the average power of the sinewave signal within a period Lf_s given by

$$P_{signal} = \frac{A^2}{2} \qquad (3.15)$$

the expected signal to the pattern noise ratio within the Nyquist rate can thus be expressed as

$$SNR_{os} = 10 \cdot \log\left(\frac{P_{signal}}{P_{os}}\right) = 10 \cdot \log\left(\frac{A^2}{2\sigma_{os}^2}\right) \qquad (3.16)$$

which shows that the SNR_{os} is decreasing at 20 dB/dec with respect to the offset errors increase. Besides, from (3.16) we can also obtain

$$\sigma_{os} = \frac{A}{\sqrt{2}} \cdot 10^{-\frac{SNR_{os}}{20}} \qquad (3.17)$$

which can be used to estimate the allowed standard deviation of offset in each parallel subfilter path. For example, for a system with 1 V_{p-p} input having total SNR_{os} greater than the mean of 40 dB (at worst case only 36 dB at 2-σ_{os} estimation), the offset standard deviation of each path must be smaller than 3.5 mV, and the mean of the noise tone will be 44 and 38 dB below the signal, respectively, according to 1- and 2-σ_{os} estimations. Although the real value could be a little bit better when taking into account the output S/H shaping effect, this result is still quite tough to reach in state-of-the-art CMOS without any specific technique especially for high-frequency operation, since not only the opamp DC offset but also SC circuit configuration, charge-injection and clock feedthrough will contribute to the path offset. For instance, considering the 4-fold 12-tap FIR interpolating filter with canonic-form ADB polyphase structure shown in Figure 2-1(a), the offset contribution for polyphase subfilter $m=0$, excluding the charge-injection and clock feedthrough, is given by

$$O_0 = (h_4 \gamma_d + h_8 \gamma_d) O_{D1} + h_8 \gamma_d O_{D2} + \gamma_{P0} O_{A1} \qquad (3.18)$$

where the O_{D1}, O_{D2} and O_{A1} are the opamp DC offset for 1st, 2nd ADB and accumulator ($m=0$), respectively, γ_d and γ_{P0} are the offset suppression factors [3.10] (all >1 for conventional SC circuits) of the SC delay (same γ_d for same delay circuit structure) and of the accumulator (γ_{P0} depends also on implemented coefficients, being different for each polyphase subfilter) circuits. Therefore, assuming the same standard deviation σ_{OA} of DC offset for all opamps, the σ_0 can be derived as

$$\sigma_0 = \sqrt{\left((h_4 + h_8)^2 + h_8{}^2\right)\gamma_d{}^2 + \gamma_{P0}{}^2} \cdot \sigma_{OA} \qquad (3.19)$$

which shows that the real path offset errors are always greater than the pure offset of the opamp. Similar procedure can be also applied to other polyphase subfilters. Note that the offsets for each path are indeed not totally independent due to the sharing of the opamp, like in ADB, so the estimation from (3.16) is not exact, but nevertheless, it is still a good prediction for the design process.

In addition, considering the non-canonic-form realization as shown in the Figure 2-1(b), the resulting offset for polyphase subfilter $m=0$ becomes

$$O_0 = h_8 \gamma_d O_{D1} + \gamma_{P0} O_{ACCU} \qquad (3.20)$$

which is not only smaller in quantity than that of the canonic-form realization but, and more importantly, the offset mismatches among the 4 pathes are significantly reduced due to the sharing of opamp for 4 path accumulation, e.g. O_{ACCU} will be a common factor for all paths and contribute mainly to the DC offset of the overall system. This obviously leads to a better performance with respect to the offset errors for the non-canonic structure when compared with the canonic. This has been verified by the simulation results for an 18-tap 4-fold FIR interpolating filter used in Chapter 3 / Session 3.1. For simplicity, 20-time Monte-Carlo simulations have been applied to both canonic (4 ADB's, 4 Accumulators, 1 MUX) and non-canonic (1 S/H, 2 ADB's, 1 Accumulator) realizations with the Gaussian random opamp offset variables with zero-mean and σ_{OA}=3.5 mV. The results are summarized in Table 3-2. The results from the canonic case match well with the theoretical estimation from (3.16). As described by (3.19) the actual offset standard deviation for each path is worst than σ_{OA}, hence the mean of SNR_{os} is a little bit worse than 40 dB obtained by using $\sigma_m = \sigma_{OA}$. In addition, it also clearly shows the consistency of the theoretical expectation implying that <u>non-canonic is superior to the canonic structure in offset sensitivity</u>. Figure 3-6 presents the pattern-noises in the output signal spectrum, from one of the cases, with non-canonic implementation. To reduce such undesired noise, offset- and gain-compensation by correlated-double sampling techniques can be employed, as will be discussed in the next chapter.

Table 3-2. Monte-Carlo Simulations of fixed pattern noise imposed by input-referred DC offset of opamps for 4-fold, 18-tap SC FIR interpolating filter (20-time, σ_{OA}=3.5 mV)

Mean (20s)	DC Offset	SNR_{os}	$SFDR_{os}$
Non-Canonic	-39 dB	56 dB	58 dB
Canonic	-41 dB	38 dB	39 dB

SFDR – Spurious-Free Dynamic Range

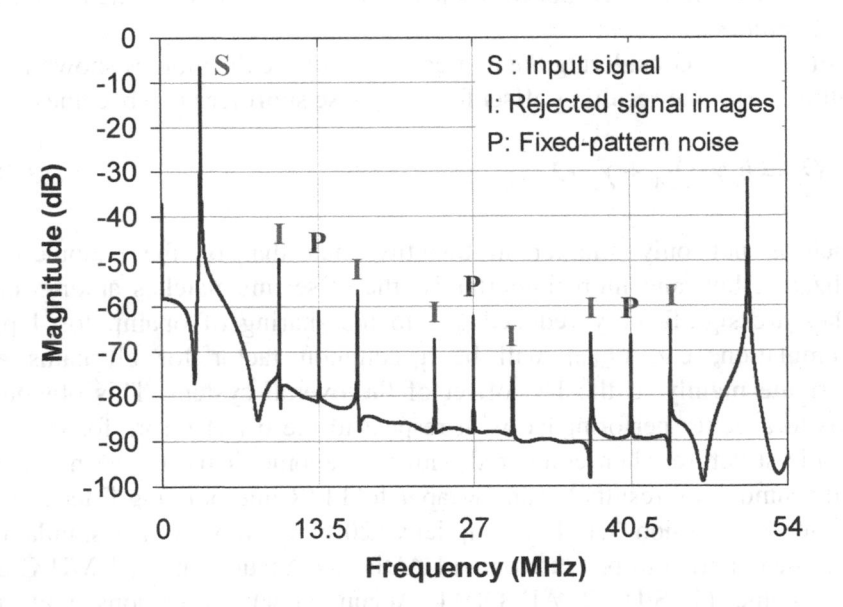

Figure 3-6. Output signal spectrum of 4-fold, 18-tap SC FIR interpolating filter (1Vp-p input, offset σ_{OA} =3.5 mV)

6. PHASE TIMING-MISMATCH EFFECTS

The parallel and multiple phase nature of multirate polyphase structures leads to the fact that the overall interpolation system suffers from the phase timing-mismatch effects that are normally unavoidable in the time-interleaved sampled-data systems. Such timing-mismatch effects can be categorized into periodic fixed timing-skew and the random timing-jitter effects.

6.1 Periodic Fixed Timing-Skew Effect

Periodic fixed timing-skew effect is mainly caused by the unmatched but periodical propagation delays among the time-interleaved phases due to systematic-design and process mismatches, as well as switching noise (dI/dt noise). The interpolation model for this type of effect can be illustrated in Figure 3-7. Timing-skew effects due to the input sampling of 4 parallel polyphase filter bands are negligible for interpolators, because the input signals are inherently sampled-and-held at the lower rate. Thus, the timing-skew errors mainly happen at the last high-speed output multiplexer stage for switching among 4 sub-filter bands at higher output rate. In opposition to the input sampling, only the rising-edge timing mismatch is the most important for output phases in order to correctly control the output signal timing.

Such fixed timing-skew renders an nonuniformly holding (in the uniform sampling input) that generates undesired modulation mirror sidebands fold back around lower input rate and its multiples within the filtering stopband that cannot be removed by the interpolating filter, while those sidebands can be shaped relatively well by the system function for the input nonuniformly sampling case. The accurate models for the nonuniformly sampling effects in the front input sampling stage (IN-OU(IS), see Appendix 1) have been well developed [3.6, 3.7, 3.8, 3.9, 3.11, 3.12]. On the other hand, for the case of interest here, with uniformly sampling input with nonuniformly playing output it has only been analyzed with ideal impulse-sampled output format (IU-ON(IS)) [3.13, 3.14]. However, the output signals are always sampled-and-held in practice for sampled-data analog interpolation, and due to this nonuniform timing, the spectrum of output signal with nonuniformly holding (IU-ON(SH)) is not just the shaped version of impulse-sampled signal spectrum, obtained by multiplying uniform $\sin x/x$ function [3.15, 3.16, 3.17, 3.18, 3.19].

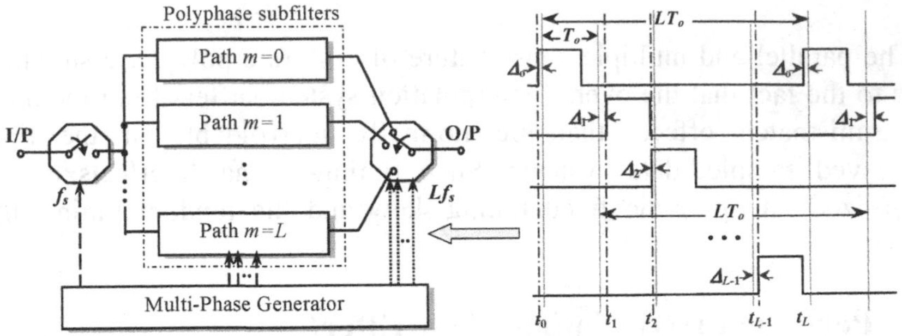

Figure 3-7. Output phase-skew sampling for polyphase-based interpolating filters

Assuming that $T_o=1/Lf_s$ is the nominal output sampling period and Δ_m is a periodic skew timing sequence with period M (path number), so that the exact output sampling instance is given by

$$t_m = nT_o + \Delta_m \qquad (3.21)$$

Let $n = kM + m$ ($m=0,1,...M$-1, M is the parallel path number that is normally equal to the interpolation factor L or has other value depending on the real circuit configuration), and the periodic skew-period ratio be $r_m = \Delta_m / T_o$, then it can be finally derived that the output signal spectrum with nonuniformly holding is

$$Y(\omega) = \frac{1}{T_o} \sum_{k=-\infty}^{\infty} A_k(\omega) \cdot X\left(\omega - k\frac{2\pi}{MT_o}\right) \qquad (3.22a)$$

where $X(\omega)$ is the input signal spectrum and

$$A_k(\omega) = \frac{1}{M} \sum_{m=0}^{M-1} H_m(\omega) e^{-jkm\frac{2\pi}{M}} e^{-j\omega r_m T_o} \qquad (3.22b)$$

$$H_m(\omega) = \frac{2\sin\left(\omega\left(1 + r_{m+1} - r_m\right)T_o/2\right)}{\omega} e^{-j\omega(1+r_{m+1}-r_m)T_o/2} \qquad (3.22c)$$

The equation (3.22) fully characterizes the output signal spectrum of the uniformly sampling and nonuniformly playing out case including the nonuniformly holding effects. Such special nonuniformly holding process causes a signal modulation at the lower input sampling rate and its multiples, i.e. mLf_s/M. Figure 3-8 presents the signal spectrum of a 58 MHz signal sampled at 320 MHz with the timing-skew effects where $M=8$ and the standard deviation of Δ_m is 5 ps. Obviously, the nonuniformly S/H output is not simply shaped by just the well-known uniform $\sin x/x$ function.

The results from the MATLAB models built according to the above equations match well with the FFT of the samples with respect to the above sampling processes. Importantly, the MATLAB models take much less number of computations than that from direct FFT. Figure 3-9 (a-c) show the mean of SNR (here only for signal to modulation sideband noise ratio) and mean of the worst noise tone to the signal, or namely Spurious-Free Dynamic Range (SFDR) within Nyquist band vs. standard deviation of the timing skew ratio r_m (skew Δ_n to output sampling period) and input signal frequencies for a parallel path number of 2, 4 and 8, respectively, from 100-time Monte-Carlo calculations. It is interesting to note that <u>both SNR and SFDR decrease at 20 dB per decade with respect to the increases of either the input signal frequencies or timing-skew errors</u>.

Furthermore, when $2\pi f_o r_m T \ll 1$, the SNR with respect to the timing jitter within the Nyquist band is derived as

$$SNR_{skew} = 20\log\left(\frac{1}{2\pi f_0 \, \sigma_t}\right) - 10\log\left(1 - \frac{1}{M}\right) \qquad (3.23)$$

where σ_t is the standard deviation of sample timing-skew error, as presented comprehensively in Appendix 1 [3.16, 3.17, 3.18]. This confirms the 20 dB/dec SNR decrease aforementioned and observed in the above simulations.

It is interesting to point out that the above SNR$_{skew}$ formula (3.23) for IU-ON(SH) is identical to the that of traditional non-uniformly impulse sampling IN-OU(IS) system with small jitter errors, i.e. $2\pi f_o r_m T \ll 1$ [3.6, 3.7, 3.8, 3.9, 3.11, 3.12]. In fact, it has also been proved in Appendix 1 that the normalized modulation sideband spectra patterns (normalizing the main sinusoidal signal component with frequency $\omega = \omega_o$ and magnitude of 1) for IN-OU(IS) and IN-OU(SH) are identical [3.16, 3.17, 3.18].

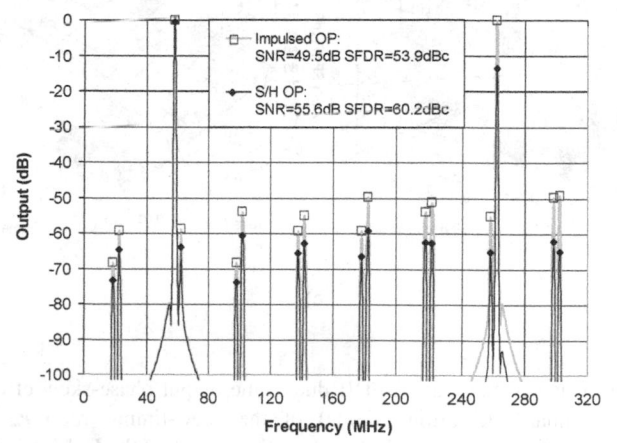

Figure 3-8. Spectrum of a 58 MHz signal sampled at 320 MHz with timing skew (M=8, σ=5 ps)

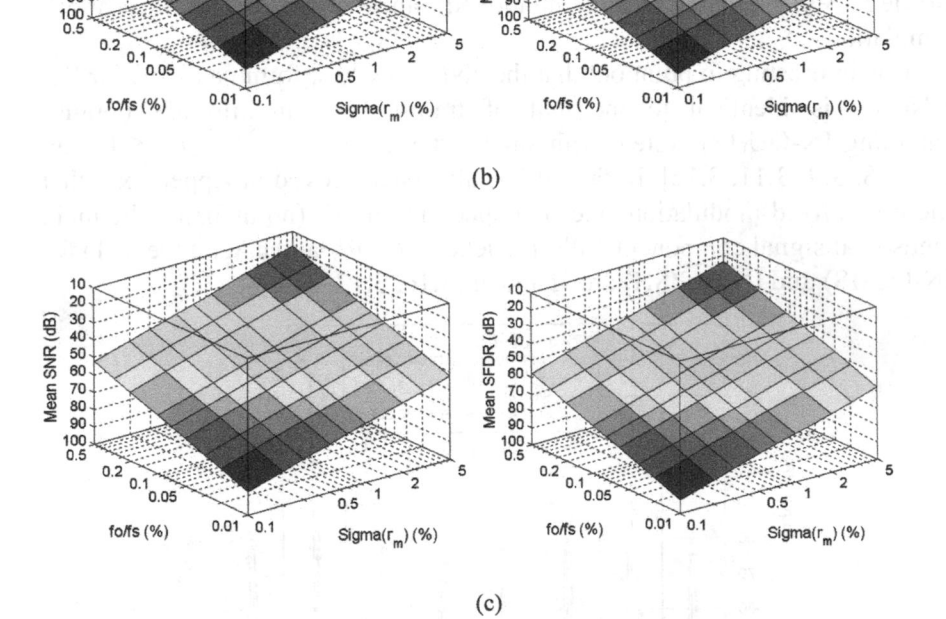

Figure 3-9. Mean value of SNR and SFDR due to the output phase-skew effects vs. signal frequencies and standard deviation (sigma) of the skew-timing ratio r_m for different interpolation factors (100-time Monte Carlo calculations) (a) *L=2* (b) *L=4* (c) *L=8*

6.2   Random Timing-Jitter Effects

Another timing-mismatch effect is caused by random timing-jitter mainly due to the device random noise that can be considered as a special case of the nonuniformly holding when path number M approaches infinity. Therefore, the equivalent SNR correlation is still valid for the nonuniformly impulse sampling and holding [3.17, 3.18]. Also, analysis of random timing-jitter effects for sampling [3.6, 3.7, 3.8, 3.20, 3.21] can be equivalently applied here where such random jitter effects result in an increased noise floor across all frequencies. For a sinewave input signal, the signal to noise ratio with sampling jitter standard deviation σ_t can be expressed as:

$$SNR_{jitter} \approx -20 \cdot \log\left(2\pi f_o \sigma_t\right) \qquad (3.24)$$

when $2\pi f_o \sigma_t \ll 1$. In addition, also a more accurate expression for this jitter error can be obtained by [3.20]

$$SNR_{jitter} = 10 \cdot \log\left(\frac{P_{signal}}{P_{jitter}}\right) = -10 \cdot \log\left(2(1 - e^{-2\pi^2 f_o^2 \sigma_t^2})\right) \qquad (3.25)$$

From the above, it can be affirmed that the resulting noise floor has 20dB/dec dependence on both the input signal frequency and jitter accuracy.

7.   NOISE ANALYSIS

The noise performance of the circuit determines the dynamic range of the SC interpolating filter. The main contributions to the total noise of this SC filter are mainly from the thermal fluctuations associated with the channel on-resistances of the MOS switches and also the transistors in the opamp. These noise sources generate mainly a broadband noise component and a sampled noise component at the output of each SC stage [3.22, 3.23, 3.24, 3.25]. The broadband noise component is originated from all the noise sources that inject it directly into the SC stage output in, at least, one phase, with cutoff frequency depending on the corresponding transients ($R_{on}C$ time constants or the opamp close-loop bandwidth) of the SC network in this phase. Since those noise sources are independent and considered uncorrelated, their contributions can be evaluated separately and simply superimposed in the end. The sampled noise component, which is normally the predominant noise contribution, is originated from the switching or sampling of the broadband noise on the capacitors. Since the noise bandwidth of the broadband noise sources are always much higher than the sampling frequency for a complete charge transfer in SC circuits, the frequency band aliasing or folding effects due to the undersampling lead to the total broadband noise being white and totally concentrated into the Nyquist band. Especially, the aliased opamp flicker noise is completely "submerged" by the aliased broadband noise of the opamp [3.22], hence, the opamp flicker noise is neglected here in the analysis that is particularly realistic for high-frequency SC filters. Besides, as mentioned before, since the offset compensation is usually preferred in an interpolating filter, the compensation will be normally associated with the elimination of the low-frequency flicker noise in nature.

A more accurate noise analysis of the SC circuits must take into account not only the bandwidth relying on the switch on-resistance and opamp frequency response but also the power transfer function from each noise source to every stage output. The latter issue leads to a very complex calculation and it is indeed impractical to analyze in a complete SC filter containing several stages where the noise sources will be generated inside the SC block of each stage. Therefore, the analysis is normally possible to be calculated for a single stage configuration, like for an SC integrator [3.22, 3.23, 3.24, 3.25, 3.26, 3.27, 3.28], an SC sample-and-hold [3.29, 3.30] and an SC gain stage [3.31, 3.32, 3.33, 3.34, 3.35]. However, in order to have an initial noise estimation of the interpolating filters, we will approximate it here only to a bandlimited noise in each stage, by switch on-resistance and opamp bandwidth during the calculation, and the practical noise shaping effects from the individual transfer function will be approximately taken into account at the last stage, considering also the shaping from the overall filter transfer function with the inherent $\sin x/x$ effects at the output.

For simplicity, the noise analysis will be concentrated on the SC interpolating filter with canonic non-recursive ADB polyphase structures. As the double-sampling is equivalent to a 2-path parallel structure, the total noise will not be doubled because the outputs are interleaved rather than summed. Therefore, only the total noise in one sampling path will be calculated, and the concept can also be applied similarly to any other structures by using the following methodology. Besides that, for a more realistic approach, the total output noise for fully-differential structures will be finally derived with the assumption also that a single-pole opamp is used.

Consequently, the total noise for the interpolating filter will be generated from ADB delay line, polyphase subfilters and output multiplexer.

A. ADB Noise Contribution: Assuming that all the SC ADB structures are the same and configured with mismatch-free property, like the second ADB in Figure 2-3, then it is presented in Figure 3-10(a) and (b) the thermal noise contributions from the switches and opamp for the ADB stage in both sampling and output phases, respectively, for only one SC sampling branch path (positive/negative half of the differential circuit).

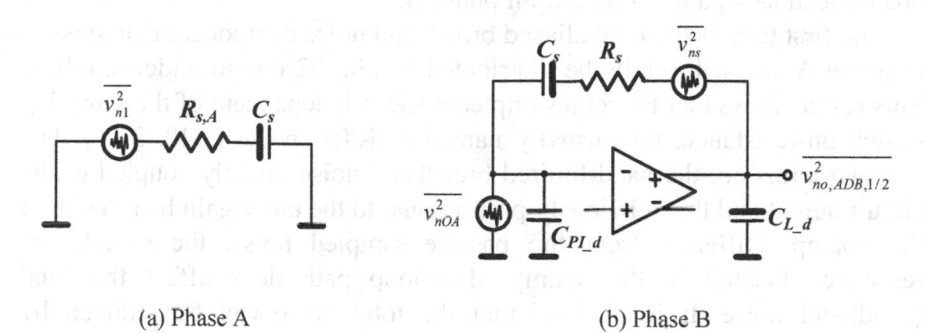

(a) Phase A   (b) Phase B

Figure 3-10. Noise in the ith mismatch-free SC ADB in (a) sampling phase A and (b) output phase B

In phase A, the broadband thermal noise due to the on-resistance is passively sampled and stored in the sampling capacitor. In phase B, both the thermal noise sources from the on-resistance and the opamp will contribute with band-limiting through its close-loop noise bandwidth. The detailed noise calculation is lengthy and it is presented in the Appendix 2. In conclusion, considering that the same opamp is used for all ADB's, the total output noise for the positive/negative half of the ADB delay line with N-number of ADBs in phase B can be obtained as

$$\overline{v^2_{no,ADB-N,1/2}} = N \cdot \overline{v^2_{no,ADB,1/2}} \qquad (3.26)$$

where the $\overline{v^2_{no,ADB}}$ is the total output-referred noise power from positive/negative half of each SC ADB obtained by the sum of all the noise contributions with the employment of Equivalent Noise Bandwidth (ENBW) [3.37, 3.38, 3.39] of a 1st-order system, i.e.,

$$\overline{v^2_{no,ADB,1/2}} = \frac{kT}{C_s} + \left[\frac{kTR_sC_s}{C_s + C_{PI\_d}} + \frac{2}{3}\frac{kT}{g_{m1\_d}}\gamma_d\left(\frac{C_s + C_{PI\_d}}{C_s}\right) \right]\omega_{ugb\_d}$$

$$(3.27)$$

where the ω_{ugb_d} and g_{m1_d} are the unity-gain bandwidth and the transconductance of the differential pair of the opamp used in the ADB stage, also γ_d represents the excess noise factor [3.36] of the same opamp that normally has a value between 1 and 3 depending on its circuit architecture. Finally, R_s is the total switch on-resistance presented in the opamp feedback path during output phase B.

The first term is the total aliased broadband noise component that appears in phase A and will totally be distributed within $f_s/2$ due to undersampling. This result shows that the total sampled noise is independent of the sampling switch on-resistance, thus usually named as KT/C noise [3.40, 3.41]. The last two terms are the band-limited broadband noise directly coupled to the circuit output and then obviously proportional to the unity-gain bandwidth of the opamp. Different from the passive sampled noise, the switch on-resistance located in the opamp close-loop path does affect the total broadband noise. It is obvious that the total noise can be reduced by increasing the sampling capacitance, reducing the opamp bandwidth, as well as the total on-resistance presented in the noise direct coupling or charge transferring path.

B. Polyphase Subfilter Noise Contribution: Figure 3-11(a) and (b) presents the noise contribution of one-path polyphase subfilter in the sampling and output summing phases. Different from the mismatch-free SC ADB, there is noise amplification due to charge transfer and thus also more thermal noise sources appear from the switch on-resistance that will contribute to the total noise in phase B. The total output noise power for positive/negative half polyphase subfilter can be finally derived as

$$\overline{v_{no,PFm,1/2}^2} = \frac{kT}{C_F^2}\sum_{i=1}^{n}C_i + \frac{kT}{C_F} + \left(\frac{kT\sum_{i=1}^{n}R_iC_i^2}{C_F(C_F + \sum_{i=1}^{n}C_i + C_{PI\_PF})}\right)\omega_{ugb\_PF}$$

$$+ \left(\frac{kTR_FC_F}{C_F + \sum_{i=1}^{n}C_i + C_{PI\_PF}}\right)\omega_{ugb\_PF} + \frac{2}{3}\frac{kT}{g_{m1\_PF}}\left(1 + \frac{g_{m7\_PF}}{g_{m1\_PF}}\right)\left(\frac{C_F + \sum_{i=1}^{n}C_i + C_{PI\_PF}}{C_F}\right)\omega_{ugb\_PF}$$

$$(3.28)$$

where C_i and R_i are the coefficient capacitors and their corresponding switch on-resistance during the charge transferring phase, R_F is the total switch on-resistance presented in the opamp feedback path during output phase B. Due to the input sampling, the KT/C noise in phase A is presented directly at the polyphase subfilter output in phase B with the coefficient gain factor, as shown in the first term of the equation (3.28). All the switch on-resistances in the charge transfer path, i.e., in input (R_i) and feedback branches (R_F) contribute to the total output noise. The last term represents the noise contribution from the opamp.

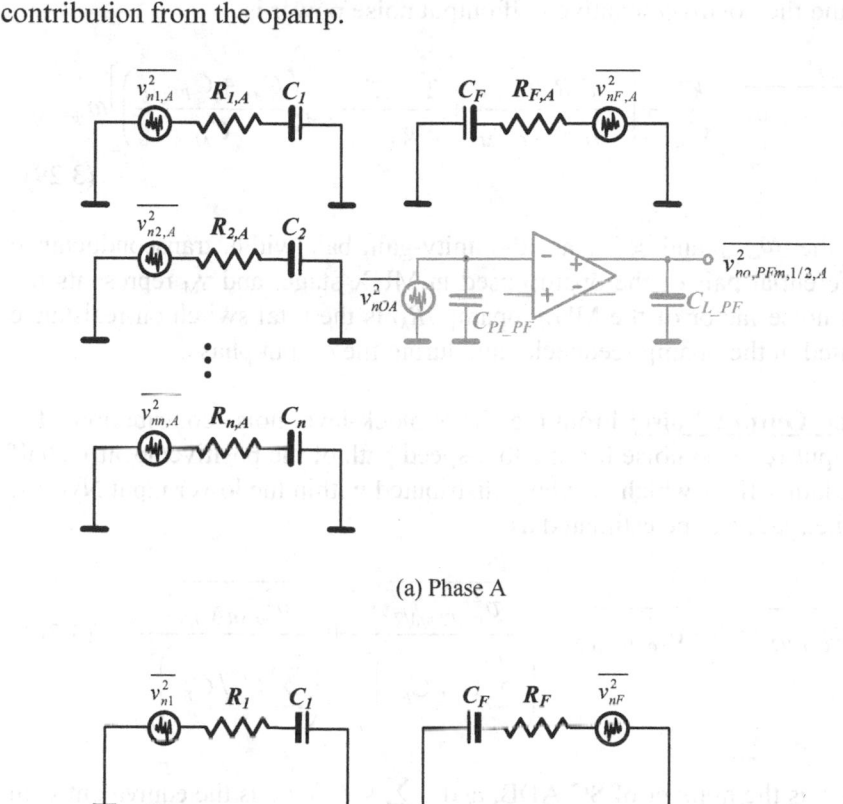

(a) Phase A

(b) Phase B

Figure 3-11. Noise in one of the L-path polyphase subfilter in (a) sampling phase A and (b) output phase B

C. Multiplexer Noise Contribution: The noise contribution for one low-speed path of the last mismatch-free output multiplexer is similar to that of the ADB so that for simplicity the noise contribution figure is not shown here, and the positive/negative half output noise power is

$$\overline{v^2_{no,MUXm,1/2}} = \frac{kT}{C_M} + \left[\frac{kTR_M C_M}{C_M + C_{PI\_M}} + \frac{2}{3}\frac{kT}{g_{m1\_M}} \gamma_M \left(\frac{C_M + C_{PI\_M}}{C_M} \right) \right] \omega_{ugb\_M}$$

(3.29)

where the ω_{ugb_M} and g_{m1_M} are the unity-gain bandwidth, transconductance of differential pair of the opamp used in MUX stage, and γ_M represents the excess noise factor of the MUX opamp. R_M is the total switch on-resistance presented in the opamp feedback path during the output phase.

D. Total Output Noise: From the above block-level noise contributions, the total input-referred noise for one low-speed path of the positive/negative half interpolating filter which is evenly distributed within the lower input Nyquist band, i.e., $f_s/2$, can be estimated as

$$\overline{v^2_{IRNm,1/2}} = N \cdot \overline{v^2_{no,ADB,1/2}} + \frac{\overline{v^2_{no,PFm,1/2}}}{\left(\sum_{i=1}^{n} C_i / C_F \right)_m^2} + \frac{\overline{v^2_{no,MUXm,1/2}}}{\left(\sum_{i=1}^{n} C_i / C_F \right)_m^2}$$

(3.30)

where N is the number of SC ADB, and $\left(\sum_{i=1}^{n} C_i / C_F \right)_m$ is the equivalent gain of each polyphase subfilter path.

Consequently, by counting the contributions from all L-path low-speed subfilters, the final total output-referred noise within the output higher Nyquist band, i.e., $Lf_s/2$, can be obtained by

$$\overline{v^2_{ORN,1/2}} = \left(\sum_{m=0}^{L-1} \overline{v^2_{IRNm,1/2}} \cdot \right) \cdot ENBW_{filtering}$$

(3.31)

where ENBW is designated as the equivalent overall filter noise bandwidth.

For the complete fully-differential SC interpolating filter, the total differential output noise can be expressed by

$$\overline{v^2_{ORN,Diff}} = 2 \times \overline{v^2_{ORN,1/2}} \qquad (3.32)$$

The above approximation methodology is very useful for an initial design of the filter noise and indeed, as it will be shown later during the design, it will match well with the measured results of the designed prototype chip.

8. SUMMARY

In this chapter, practical IC design issues for the previously proposed multirate polyphase-based structures have been investigated comprehensively in terms of the power dissipation and circuit achievable accuracy with respect to the capacitor ratio mismatches, finite-gain & bandwidth, offset effects, timing-mismatch effects as well as noise analysis.

FIR interpolation with the non-recursive ADB polyphase structures shows their superiority in very-low-sensitivity and inherent linear-phase characteristics in the passband. A more accurate approximation to worst-case stopband magnitude deviation has also been derived. And, from this estimation it can be predicted that FIR interpolation is very suitable for high-frequency video analog front-end filtering with 7-8 bit accuracy and linear-phase requirements.

Among several architectures based on Recursive-ADB polyphase structures with combined non-recursive ADB structure and direct-form II recursive networks for IIR interpolation, Extra-ripple transfer functions are shown to be especially appropriate due to their added value in lowering the sensitivity in the passband, but with similar level in the stopband when compared to conventional IIR transfer functions, like Elliptic, Chebyshev approximations. The proposed R-ADB with Mixed Cascade/Parallel-form DFII structure has the lowest sensitivity over the R-ADB with either Complete-DFII or Parallel-DFII, especially for higher-order filtering.

Canonic-form realizations of the ADB polyphase structures have been specialized for very high-frequency applications with the employment of L-output-accumulators plus double-sampling techniques. While the non-canonic-from realization which uses one-output-accumulator and requires less number of ADB's presents its superiority in reduced component counts and circuit sensitivity, thus being more suitable for relatively lower-frequency with higher-accuracy applications. Especially, the power analysis shows also that the power of canonic-form realizations could be still lower than the non-canonic even though the number of opamps and SC branches in the former are more than the double of those in the latter.

Furthermore, the practical input-referred DC offset and phase timing-skew & random jitter effects have also been analyzed comprehensively for interpolation circuits with also the concise signal-to-noise expectations. Such imperfections imply that specific techniques are mandatory for higher performance, especially in high-frequency applications.

REFERENCES

[3.1] Z.Q.Shang, J.I.Sewell, "Development of efficient switched network and mixed-mode simulator," *IEE Proc. Circuits, Devices and Systems*, Vol.145, No.1, pp.24-34, Feb. 1998.

[3.2] A.Petraglia, S.K.Mitra, "Effects of coefficient inaccuracy in switched-capacitor transversal filters," *IEEE Trans. Circuits and Systems*, Vol.38, No.9, pp.977-983, Sep. 1991.

[3.3] A.Petraglia, "Fundamental frequency response bounds of direct-form recursive switched-capacitor filters with capacitance mismatch," *IEEE Trans. Circuits and Systems – II: Analog and Digital Signal Processing*, Vol.48, No.4, pp.340-350, Apr. 2001.

[3.4] A.Petraglia, *Mixed Analog/Digital Structures for High-Speed A/D Conversion and Signal Processing*, Ph.D. Dissertation, University of California, Santa Barbara, USA, 1991.

[3.5] A.Petraglia, S.K.Mitra, "Analysis of mismatch Effects among A/D converters in a time-interleaved waveform digitizer," *IEEE Trans. Instrumentation and Measurement*, Vol.40, No.5, pp.831-835, Oct. 1991.

[3.6] Y.C. Jenq, "Digital Spectra of Nonuniformly Sampled Signals: Fundamentals and High-Speed Waveform Digitizers," *IEEE Trans. Instrumentation and Measurement*, vol.37, no.2, pp.245–251, Jun.1988.

[3.7] G.T.Uehara, *Circuit techniques and considerations for implementation of high speed CMOS analog-to-digital interfaces for DSP-based PRML magnetic disk read channels*, Ph.D. Dissertation, University of California, Berkeley, USA, 1993.

[3.8] M.Gustavssn, *CMOS A/D converters for telecommunications*, Ph.D. Dissertation, Linköping Universitet, Sweden, 1998.

[3.9] N.Kurosawa, H.Kobayashi, K.Maruyama, et al., "Explicit Analysis of Channel Mismatch Effects in Time-Interleaved ADC Systems," *IEEE Trans. Circuits and Systems – I*, vol.48, No.3, pp.261-271, Mar.2001.

[3.10] Wing-Hung Ki, G.C.Temes, "Gain- and Offset-compensated Switched-Capacitor filters," in *Proc. IEEE International Symposium on Circuits and Systems (ISCAS)*, pp.1561-15664, 1991.

[3.11] A.Yu, W.C.Black, Jr., "Error Analysis for time-interleaved analog channels," in *Proc. IEEE International Symposium on Circuits and Systems (ISCAS)*, Vol.I, pp.468-471, May 2001.

[3.12] M.Gustavsson, N.N.Tan, "A global passive sampling technique for high-speed Switched-Capacitor time-interleaved ADCs," *IEEE Trans. Circuits and Systems – II: Analog and Digital Signal Processing*, vol.47, No.9, pp.821-831, Sep.2000.

[3.13] Y.C.Jenq, "Digital-To-Analog (D/A) converters with nonuniformly sampled signals," *IEEE Trans. Instrumentation and Measurement*, vol.45, No.1, pp.56-59, Feb.1996.

[3.14] Y.C.Jenq, "Direct digital synthesizer with jittered clock," *IEEE Trans. Instrumentation and Measurement*, vol.46, No.3, pp.653-655, Jun. 1997.

[3.15] Seng-Pan U, R.P.Martins, J.E.Franca, "Design and analysis of low timing-skew clock generation for time-interleaved sampled-data systems," *Proc. IEEE International Symposium on Circuits and Systems (ISCAS)*, pp.441-444, May. 2002.

[3.16] Sai-Weng Sin, Seng-Pan U, and R.P.Martins, "Timing-Mismatch Analysis in High-Speed Analog Front-End with Nonuniformly Holding Output," *Proc. IEEE International Symposium on Circuits and Systems (ISCAS)*, pp.I-129 - I-132 vol.1, May 2003.

[3.17] Seng-Pan U, Sai-Weng Sin and R.P.Martins, "Spectra Analysis of Nonuniformly Holding Signals for Time-Interleaved Systems with Timing Mismatches," *Proc. IEEE Instrumentation and Measurement Technology Conference - IMTC'2003*, vol. 2, pp. 1298-1301, May 2003.

[3.18] Seng-Pan U, Sai-Weng Sin and R. P. Martins, "Exact Spectra Analysis of Sampled Signals with Jitter-Induced Nonuniformly Holding Effects," *IEEE Transactions on Instrumentation and Measurement*, vol. 53, pp. 1279-1299, Aug. 2004.

[3.19] Sai-Weng Sin, Seng-Pan U and R.P.Martins, "Quantitative Noise Analysis of Jitter-Induced Non-Uniformly Sampled-And-Held Signals," in *Proc. IEEE International Conference on Acoustics, Speech and Signal Processing – "ICASSP'2003"*, vol. 6, pp. VI-253 - 256, April 2003.

[3.20] H.Kobayashi, M.Morimura, et al., "Aperture jitter effects in wideband sampling systems," in *Proc. of IEEE Instrumentation and Measurement Technology Conference – IMTC'99*, pp.880-885, May 1999.

[3.21] J.L.González, E.Alarcón, "Clock-jitter induced distortion in high speed CMOS Switched-Current segmented digital-to-analog converters," in *Proc. IEEE International Symposium on Circuits and Systems (ISCAS)*, pp.512-515, May 2001.

[3.22] C.A.Gobet, A.Knob, "Noise analysis of Switched Capacitor networks," *IEEE Trans. Circuits and Systems*, vol.CAS-30, No.1, pp.37-43, Jan.1983.

[3.23] R.Castello, P.R.Gray, "Performance limitations in Switched-Capacitor filters," *IEEE Trans. Circuits and Systems*, vol.CAS-32, No.9, pp.865-876, Sep.1985.

[3.24] R.Gregorian, G.C.Temes, *Analog MOS Integrated Circuits for Signal Processing*, John Wiley & Sons, Inc., 1986.

[3.25] J.Goette, W.Guggenbühl, "Noise performance of SC-Integrators assuming different operational transconductance amplifier (OTA) models," *IEEE Trans. Circuits and Systems*, vol.35, No.8, pp.1042-1048, Aug.1988.

[3.26] A.K.Ong, *Bandpass Analog-to-Digital Conversion for Wireless Applications*, Ph.D. Dissertation, Standford University, USA, 1998.

[3.27] J.Grilo, *Improved Design Techniques for Low-Voltage Low-Power Switched-Capacitor Delta-Sigma Modulator*, Ph.D. dissertation, Oregon State University, USA, 1997.

[3.28] A.Marques, *High-speed CMOS data converters*, Ph.D. dissertation, Katholieke Universiteit Leuven, Belgium, 1999.

[3.29] R.A.Gomez, *A Discrete-Time Analog Read Channel IC for Magnetic Recording*, Ph.D. Dissertation, University of California, Los Angeles, USA, 1993.

[3.30] D.H.Shen, *Architecture and Design of a Monolithic Radio Frequency Receiver*, Ph.D. Dissertation, Stanford University, USA, 1997.

[3.31] T.Cho, *Low-Power Low-Voltage Analog-to-Digital Conversion Techniques using Pipelined Architectures*, Ph.D. Dissertation, University of California, Berkeley, USA 1995.

[3.32] K.Y.Kim, *A 10-bit, 100MS/s Analog-to-Digital Converter in 1-μm CMOS*, Ph.D. Dissertation, University of California, Los Angeles, USA, 1996.

[3.33] A.M.Abo, *Design for Reliability of Low-voltage, Switched-Capacitor Circuits*, Ph.D. Dissertation, University of California, Berkeley, USA, 1999.

[3.34] J.Goes, *Optimization of Self-Calibrated CMOS Pipelined Analogue-to-Digital Converters*, Ph.D. Dissertation, Instituto Superior Técnico, Portugal, 2000.

[3.35] M.Gustavsson, J.J.Wikner, N.N.Tan, *CMOS Data Converters for Communications*, Kluwer Academic Publishers, 2000.

[3.36] K.R.Laker, W.M.C.Sansen, *Design of Analog Integrated Circuits and Systems*, McGraw-Hill, Inc., 1994.

[3.37] D.A.Johns, K.Martin, *Analog Integrated Circuit Design*, John Wiley & Sons, Inc., 1997.

[3.38] P.R.Gray, P.J.Hurst, S.H.Lewis, R.G.Meyer, *Analysis and Design of Analog Integrated Circuits*, 4th-edition, John Wiley & Sons, Inc., 2001.

[3.39] A.Razavi, *Design of Analog CMOS Integrated Circuits*, McGraw-Hill, Inc., 2001.

[3.40] R.J.Baker, H.W.Li, D.E.Boyce, *CMOS Circuit Design, Layout, and Simulation*, IEEE Press, 1997.

[3.41] P.E.Allen, D.R.Holberg, *CMOS Analog Circuit Design*, 2nd-edition, Oxford University Press, Inc., 2002.

Chapter 4

GAIN- AND OFFSET- COMPENSATION FOR MULTIRATE SC CIRCUITS

1. INTRODUCTION

The imperfections of the active element – opamp in terms of the finite gain, bandwidth and DC offset are the key issues to take into consideration in the design of high-performance multirate SC filters, especially for high-frequency operation. Both finite gain and bandwidth lead to variations in the system response and increased nonlinearity. Nevertheless, the large bandwidth of the opamp trades off with the open-loop gain. Moreover, as the application- and technology-driven constraints are scaling down the system supply voltage, large output signal swing also trades off with the achievable opamp's gain. Furthermore, the DC offsets of ADB's, which will be accumulated along the serial delay line, and that of the accumulators will both contribute for the offset mismatches among parallel polyphase subfilters, thus rendering undesired fixed-pattern noise. From the analysis in the previous chapter, this effect becomes problematic and reduces significantly the SNR in conventional SC implementations imposing the mandatory need for specific circuit techniques to ensure higher performance in the system.

The Autozeroing (AZ) and Correlated Double Sampling (CDS) techniques, which will both sample and subtract the amplifier noise and offset in each clock period, are inherently appropriate for sampled data circuits to compensate the opamp imperfections, i.e., 1/f noise and DC offset effects, where CDS could also achieve extra functionality by lowering the sensitivity to the finite gain [4.1]. This chapter will propose novel circuit architectures and building blocks for high performance multirate SC circuits with the employment of both AZ and CDS techniques. After a brief study on the AZ and various CDS techniques, different novel Gain- and Offset Compensated (GOC) SC delay circuit cells will first be proposed with added superiority of Mismatch-Free (MF) properties. Secondly, different GOC SC accumulation circuitries will be proposed. Specific design examples will also be built using the proposed building blocks. Finally, practical implementation issues in terms of AC analysis of the GOC circuits, which is one of the important considerations for high-frequency operation, will be investigated and discussed.

2.   AUTOZEROING AND CORRELATED-DOUBLE SAMPLING TECHNIQUES

Autozeroing (AZ) and Correlated-Double Sampling (CDS) techniques [4.1, 4.2] both require two consecutive phases for operation: during the first sampling or calibration phase, the non-ideal virtual ground voltage, or error voltage of opamp as shown in Figure 4-1, i.e. input-referred DC offset, finite-gain and low-frequency noise (1/f noise), is sampled and stored across capacitors; during the next operation or compensation phase (when output is being sampled), this error voltage will be subtracted from the signal voltage by appropriate switching of the capacitors. This subtraction is attributed by the highpass filtering to the error noise due to the introduced zero at the origin in the baseband power transfer function, and the resulting error compensation will also improve the linearity of the circuit [4.1, 4.3].

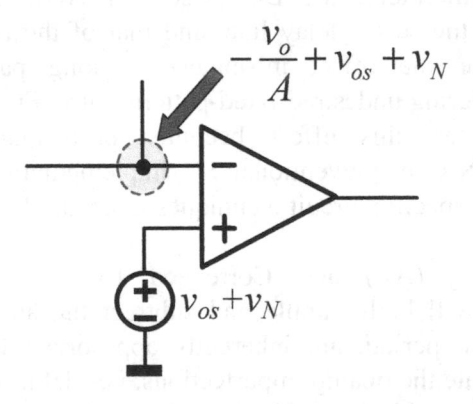

Figure 4-1. Virtual ground error voltage compensated by AZ or CDS techniques

In AZ circuits, the opamp needs to be reset to zero ($v_o=0$) during the calibration phase, therefore, only the offset errors and low-frequency 1/f noise will be stored by the capacitors and can be suppressed in the next phase. Differently, for CDS, opamp settles during calibration phase at a certain voltage level correlated to the next operation output voltage, thus, such information will help in an extra compensation of the opamp finite-gain errors.

According to the correlation intensity to the output voltages in calibration and operation output phases, CDS can be divided into two categories: Holding-CDS (H-CDS) [4.3, 4.4, 4.5, 4.6, 4.7, 4.8, 4.9, 4.10, 4.11] and Predictive-CDS (P-CDS) [4.12, 4.13, 4.14, 4.15, 4.16, 4.17, 4.18, 4.19]. H-CDS holds the previous-phase output voltage while the P-CDS tends to precisely anticipate the future value of the output voltage for the calibration use. Thus, H-CDS can effectively boost the gain of the opamp only in the narrow baseband where the signals don't vary much from one clock phase to the next, like in the oversampling applications; while the P-CDS is able to compensate the finite-gain error very precisely over a very wide signal band but with the price of additional circuitry. This is mainly due to the essential duplication of the SC branches originally used in the circuit for performing an anticipatory process of obtaining the virtual ground voltage that is expected to be present during the next compensation phase.

Basically, two kinds of methods can be used for storing the error voltage at the virtual ground: first directly by using the input Sampling Capacitor (SC) [4.5, 4.6, 4.10, 4.12, 4.16] that stores input signal sample and virtual ground errors simultaneously, while another by adding an extra capacitor, namely Error-storage Capacitor (EC) [4.4, 4.8, 4.9, 4.10, 4.11, 4.13, 4.14, 4.15, 4.16, 4.17, 4.18, 4.19] connected to the virtual ground for solely storing the virtual ground error voltage and then generating a "super virtual ground" in series with the input branch during the next output phase. For simplicity, we designated the former and latter CDS schemes as SC/CDS and EC/CDS, and by applying this to the previous categories it leads to SC/H-CDS or SC/P-CDS and EC/H-CDS or EC/P-CDS, respectively, as shown in Figure 4-2. Since in the SC/CDS scheme, the sampling capacitor is connected between the input signal terminal and virtual ground directly, the circuit output will vary due to the direct charge coupling between this sampling and feedback capacitors that will interfere with the virtual ground voltage, hence, in other words, affecting the error compensation accuracy. Besides, this charge coupling also restricts the correct output timing not to happen at input sampling phase. While in the latter EC/CDS approach, the input signal sampling and virtual ground error storage will be done by two different capacitors that make such two processes independent, and hence, a more accurate compensation can be normally achieved. However, the regenerated "super virtual ground" is quite sensitive to parasitic capacitance and especially the addition of the extra error-storage capacitor will increase the effective close-loop capacitance loading for the opamp which limits the achievable speed of the overall circuit as it will be discussed later.

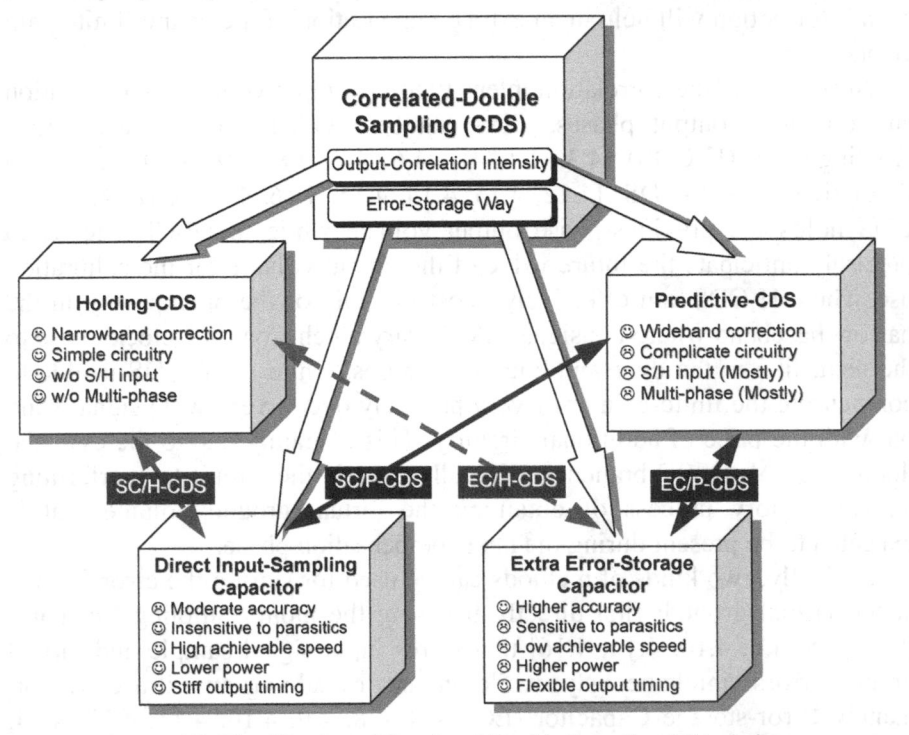

Figure 4-2. Classification of Correlated-Double Sampling SC techniques

3. AZ AND CDS SC DELAY BLOCKS WITH MISMATCH-FREE PROPERTY

3.1 SC Delay Block Architectures

As the SC unit-delay ADB's are cascaded as a serial processing delay line, which contributes to the transfer function coefficients implementation, their errors will be accumulated along this serial delay line that will considerably affect the overall system performance, e.g. finite gain errors lead to the reduction in filter coefficient precision and system linearity; DC offsets imposes fixed pattern noise. In addition, for further improvement of accuracy, the elimination of the mismatches among the capacitance ratios is also indispensable.

Different SC structures can be used as delay cell [4.16, 4.18, 4.20, 4.21, 4.22, 4.23, 4.24, 4.25, 4.26], here, as depicted in Figure 4-3, we propose several novel implementations of GOC and Mismatch-Free (MF) SC delay circuits using CDS technique [4.27], together with a typical conventional UnCompensated (UC) SC delay circuit of Figure 4-3(a) for reference and comparison. Figure 4-3(b) presents uncompensated SC delay circuit with mismatch-free property that was used as ADB in the previous chapter. Since there is no reset phase in the circuit, this means that half-unit or full-unit delays can be achieved in practice. Figure 4-3(c) is an AZ MF half-period delay circuit, and it is the simplest one among all other circuits where only one reset phase is needed for calibrating the offset and low-frequency errors. It is also evident that the input sampled and the output signals of circuits Figure 4-3(b)-(j) are all represented by the same charges on the same capacitor, which implies that no charge transferring operation exists, thus eliminating physical mismatch problem of capacitance ratio.

The novel circuits proposed in Figure 4-3(d) and (e) implement SC/H-CDS and EC/H-CDS, respectively, for achieving both gain- and offset-compensation. As we can see, both of these two circuits realize the H-CDS using feedback path of C_F which holds previous phase 2 output voltage for generating the nonzero inverting node voltage in phase 1 that is stored either in sampling capacitor or extra error-storage capacitor. This stored voltage will then be used for approximately compensating the gain error in the next operation or output phase 2. It is obvious that the drawback occurs when the input is not oversampled (meaning that the output varies much from one phase to the next) since the finite gain effect is increased extensively and cannot be compensated accurately, which implies a frequency-dependent compensation. Especially, the sampling capacitor C_1 (circuit in Figure 4-3(d)) will directly couple certain amount of charge to C_F depending on the variation of input signals between the two phases, during the calibration phase, thus affecting the virtual ground node voltage. While for the circuit in Figure 4-3(e) [4.28], by having a fixed charge redistribution between the error-storage C_h and C_F, a so-called "super virtual ground" at node x is generated during the operation/output phase 2, thus leading to better performance in overall compensation.

The circuits of Figure 4-3(f) and (g) correspond to a further enhancement of the circuits of Figure 4-3(d) and (e) respectively for fulfilling the SC/H-CDS and EC/H-CDS schemes. By performing a similar delay switching operation preliminarily, i.e. by a duplicate SC branch with C_F, a much closer approximation of finite gain and offset errors that are expected to be presented in next output phase can be obtained, thus attaining a very accurate and almost frequency-independent compensation. Similarly as in Figure 4-3(d), the charge coupling between C_1 and C_F will degrade the compensation performance for circuit of Figure 4-3(f). In addition, all these two circuits require an extra phase for connecting them with previous stage due to the fact that the input must be held constant over two phases, because one phase is necessary for sampling from original output SC branch and another for the duplicate SC branch phase of the prediction process.

The aforementioned input S/H condition and also the duplication of SC branches for prediction operation are no longer required in the following two circuits of Figure 4-3(h) and (i) that perform EC/P-CDS in a different way at the price of extra phases. The same circuit configuration of Figure 4-3(e) is employed to achieve the anticipatory compensation successively with multiple phases in a two-step approach, as shown in Figure 4-3(h). The circuit samples the input only once and switches the sampling capacitor C_1 twice, where in the first time, the circuit performs H-CDS in phase 1 and 1' while in the second time fulfills P-CDS in phase 2 and 2'. It is also possible not to increase the speed requirement of the opamp, which is especially appropriate for sampling rate converter circuits due to their multirate nature. Also the Figure 4-3(i) implements the anticipatory compensation process using the Same Sample Correction (SSC) property as referred in [4.14] by switching the virtual ground from un-compensated (calibration phase) to error-compensated (output operation phase) by the switching of the error-storage capacitor in the same output phase. By elegant configuration, this SSC technique can be applied to any conventional SC circuits to achieve CDS property by just inserting an SSC branch between input sampling branches and opamp input as well as adding two operational phases, i.e. E & O, as can be seen from Figure 4-3(b) and (i).

Finally, Figure 4-3(j) shows the implementation of also a novel MF delay circuit with EC/P-CDS property which is particularly suitable for fully-differential circuit which is a dominant technique for state-of-the-art low-voltage SC circuits due to its superiority in noise and output swing as well as speed performance. In phase 1, the circuit performs at the same time signal sampling from positive input at C_1 and direct signal coupling from the negative input to the output through C_2 and C_F where the resulting finite-gain and offset error voltages reflected to the virtual ground are also stored in C_h. Thus, if $C_2/C_F =1$, the circuit can accomplish also an accurate P-CDS compensation. Especially, the circuit only needs two phases and more importantly no input S/H requirement is required.

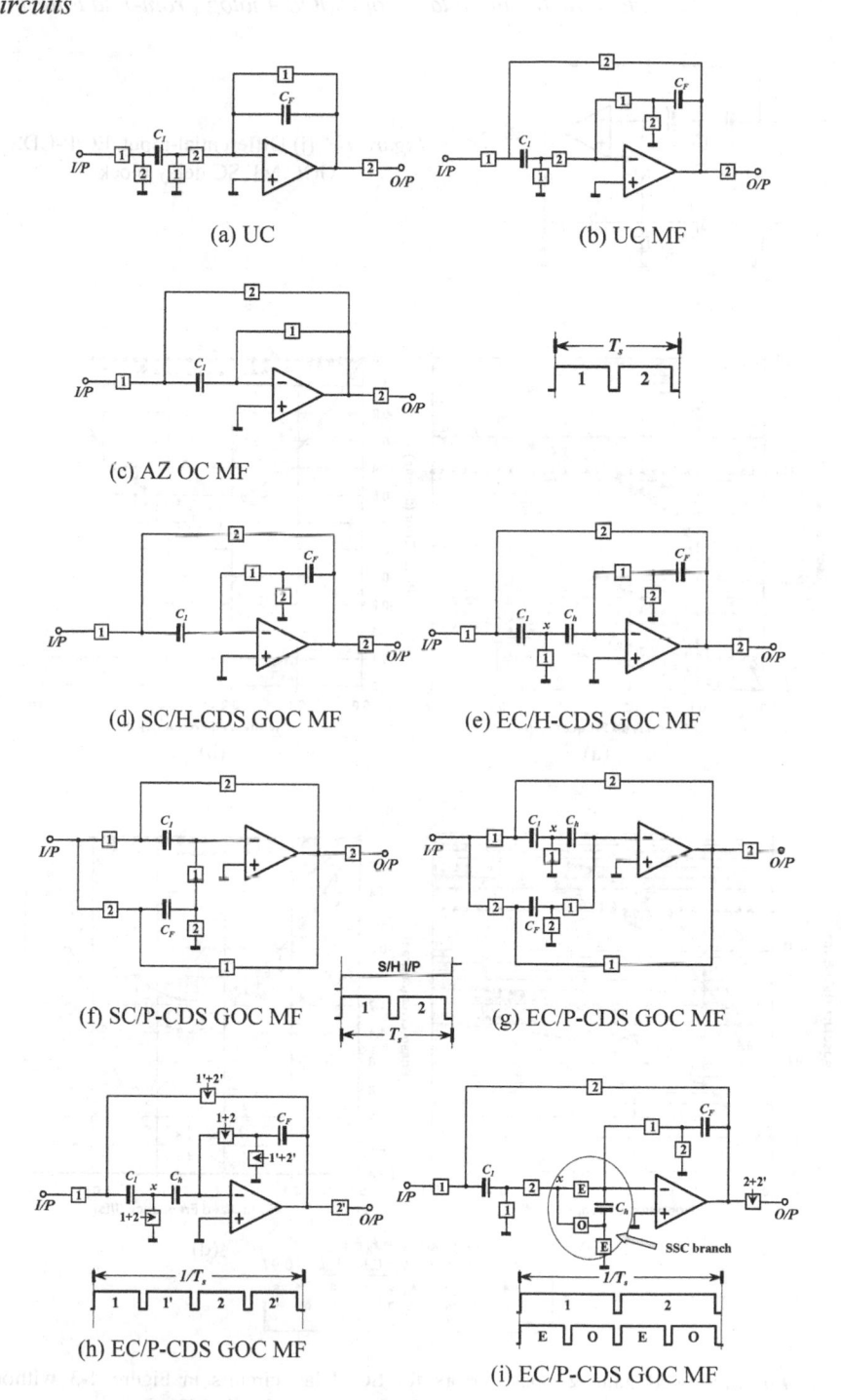

Figure 4-3. Different mismatch-free SC delay blocks with UC, AZ and CDS techniques

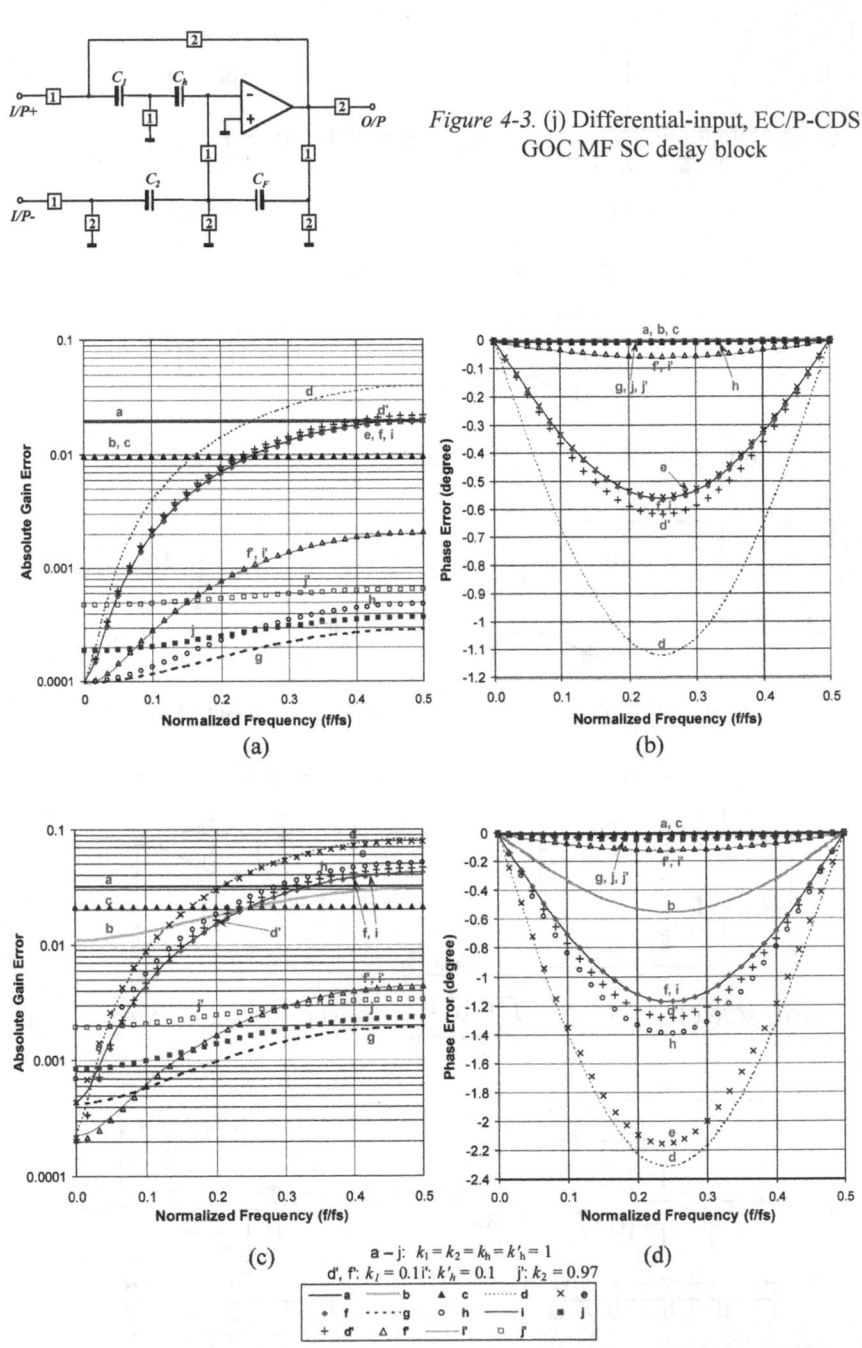

Figure 4-3. (j) Differential-input, EC/P-CDS GOC MF SC delay block

a – j: $k_1 = k_2 = k_h = k'_h = 1$
d', f': $k_I = 0.1$ i': $k'_h = 0.1$ j': $k_2 = 0.97$

Figure 4-4. Simulated gain & phase errors for SC delay circuits in Figure 4-3 without parasitics (a) & (b) and with parasitics (c) & (d) (Parasitics: 10% & 30% @ capacitor top & bottom plate, C_p@opamp input node $=C_F$)

3.2 Gain and Offset Errors – Expressions and Simulation Verification

To facilitate the comparison of all these delay circuits' performance in terms of finite gain compensation, the relative magnitude and phase errors of the overall frequency response will be evaluated. The non-ideal frequency response due to the finite gain can be approximated by [4.29]

$$H(j\omega) \approx \frac{H_D(j\omega)}{1 - m(j\omega) - j\theta(j\omega)}, \text{ when } m(j\omega), \theta(j\omega) << 1 \qquad (4.1)$$

where the $H_D(j\omega)$ is the ideal response of the SC delay circuit while $m(j\omega)$ and $\theta(j\omega)$ are its relative gain and phase errors, respectively. In addition, to explore the effect from the DC offset of opamp to the circuit, the γ-factor, namely offset suppression factor [4.9] is employed here, so that the total offset error of the circuit will be γ times opamp DC offset voltage v_{os}. We have derived rigorously the closed-form expressions for these errors for all of the above circuits that are summarized in Table 4-1, where A is the finite gain of the opamp.

From Table 4-1, it is clearly shown that the gain error of the UC SC delay circuit with MF property, e.g. Figure 4-3(b) & (c), is always $1/A$ and normally half of that of the conventional UC SC delay circuit that needs charge transferring like in Figure 4-3(a). Note that, for circuit of Figure 4-3(b), both the gain and offset errors will be doubled in the output holding process at phase 2 that could also be used as a delayed signal.

Except for the AZ SC circuit of Figure 4-3(c), where the gain error is frequency-independent and kept constant at $1/A$, all the CDS circuits have their gain and phase errors dependent of the frequency and described respectively by cosine and sine function with different weighted magnitude. As it can be seen, due to the H-CDS nature aforementioned, circuits like the one in Figure 4-3(d) & (e), have weight factor of $1/A$ while the P-CDS circuits, e.g. Figure 4-3(g), (h) and (j), have a factor of $1/A^2$, so that the latter is much more insensitive to the frequency variation. Note that although circuits of Figure 4-3(f) and (i) tend to operate in P-CDS mode, the corresponding direct charge-coupling effect between C_1 and C_F for (f) and charge-redistribution between C_h and C_1 for (i) will both disturb virtual ground voltage or indeed the prediction precision, thus the corresponding weighted factor is $1/A$ times their ratios, i.e. $k_1 = C_1 / C_F$ and $k'_h = C_h / C_1$. Hence, it can also be deduced that the performance of those two will be improved by letting k_1 and k'_h be smaller.

78

Design of Very High-Frequency Multirate Switched-Capacitor Circuits –
Extending the Boundaries of CMOS Analog Front-End Filtering

Table 4-1. Gain & phase errors and offset-suppression factor for SC delay circuits in Figure 4-3 (a)-(j)

Figure 4-3	Gain Error $m(j\omega)$	Phase Error $\theta(j\omega)$	Offset Suppression Factor γ
(a)	$-(1+k_1)\mu$	0	$\dfrac{(1+k_1)}{1+\mu(1+k_1)}$
(b)	$-\mu \; ; \; (-2\mu-\mu^2 \text{ @ phase 1})$	0	$\dfrac{1}{1+\mu} \; ; \; \left(\dfrac{2+\mu}{(1+\mu)^2} \text{ @ phase 1}\right)$
(c)	$-\mu$	0	$\dfrac{\mu}{(1+\mu)^2}$
(d)	$-(1+k_1)\mu + \dfrac{(1+k_1)\mu + k_h\mu^2}{1+\mu}\cos(\omega T_s)$	$-\dfrac{(1+k_1)\mu + k_h\mu^2}{1+\mu}\sin(\omega T_s)$	$\dfrac{\mu}{(1+\mu)(1+(1+k_1)\mu)}$
(e)	$-\mu + \dfrac{\mu}{1+(1+k_h)\mu}\cos(\omega T_s) + \dfrac{k_h\mu^2}{(1+(1+k_h)\mu)^2}\cos(2\omega T_s)$	$-\dfrac{\mu}{1+(1+k_h)\mu}\sin(\omega T_s) + \dfrac{k_h\mu^2}{(1+(1+k_h)\mu)^2}\sin(2\omega T_s)$	$\dfrac{\mu}{(1+\mu)(1+(1+k_h)\mu)} + \dfrac{k_h\mu^2}{(1+\mu)(1+(1+k_h)\mu)^2}$
(f)	$-\dfrac{\mu^2 + k_h\mu(1+\mu)(1-\cos(\omega T_s))}{1+2\mu}$	$-\dfrac{k_h\mu(1+\mu)\sin(\omega T_s)}{1+2\mu}$	$\dfrac{\mu}{(1+\mu)(1+(1+k_1)\mu)}$
(g)	$-\dfrac{(1+k_h)\mu^2}{1+(2+k_h)\mu} + \dfrac{k_h\mu^2(1+\mu)(1+(1+k_h)\mu)(1+(2+k_h)\mu)\cos(\omega T_s)}{(1+(1+k_h)\mu)(1+(2+k_h)\mu)+2k_h\mu^2\cos(\omega T_s)}$	$-\dfrac{k_h\mu^2(1+\mu)(1+(1+k_h)\mu)(1+(2+k_h)\mu)\sin(\omega T_s)}{(1+(1+k_h)\mu)(1+(2+k_h)\mu)+2k_h\mu^2\cos(\omega T_s)}$	$\dfrac{\mu}{(1+\mu)(1+(1+k_h)\mu)} + \dfrac{k_h\mu^2}{(1+\mu)(1+(1+k_h)\mu)^2}$
(h)	$-\dfrac{(2+k_h)\mu^2+(1+k_h)\mu^3}{1+(3+k_h)\mu+(1+k_h)\mu^2} + \dfrac{((1+k_h)\mu^2+k_h\mu^3)\cos(\omega T_s)}{(1+\mu)(1+(1+k_h)\mu)(1+(3+k_h)\mu+(1+k_h)\mu^2)}$	$\dfrac{((1+k_h)\mu^2+k_h\mu^3)\sin(\omega T_s)}{(1+\mu)(1+(1+k_h)\mu)(1+(3+k_h)\mu+(1+k_h)\mu^2)}$	$\dfrac{\mu^2}{(1+\mu)(1+(1+k_h)\mu)} + \dfrac{\mu^2}{(1+\mu)^2(1+(1+k_h)\mu)^2} + \dfrac{k_h\mu^2}{(1+\mu)(1+(1+k_h)\mu)^2}$
(i)	$-\dfrac{(k'_h+(1+k'_h)\mu)\mu}{1+2\mu} + \dfrac{k'_h\mu}{1+(1+k_h)\mu}\cos(\omega T_s)$	$-\dfrac{k'_h\mu}{1+(1+k_h)\mu}\sin(\omega T_s)$	$\dfrac{k'_h(1+2\mu)\mu+(1+2\mu)-(1+(1+k'_h)\mu)(1+(1+k_h)\mu)}{(1+\mu)(1+(1+k'_h)\mu)(1+(1+k_h)\mu)} + \dfrac{k_h\mu^2}{(1+\mu)(1+(1+k_h)\mu)^2}$
(j)	$-\dfrac{\mu(1-k_2+\mu(1+k_h+k_2))}{1+\mu(1+k_h+2k_2)} + \dfrac{\mu^2 k_h k_2\cos(\omega T_s)}{(1+\mu(1+k_h+2k_2))(1+\mu(1+k_h+k_2))}$	$-\dfrac{\mu^2 k_h k_2\sin(\omega T_s)}{(1+\mu(1+k_h+2k_2))(1+\mu(1+k_h+k_2))}$	$\dfrac{\mu(1+k_2)(1+\mu(1+2k_h+k_2))}{(1+\mu)(1+\mu(1+k_h+k_2))^2}$

where $\mu = \dfrac{1}{A}$, $k_1 = \dfrac{C_1}{C_F}$, $k_2 = \dfrac{C_2}{C_F}$, $k_h = \dfrac{C_h}{C_F}$, $k'_h = \dfrac{C_h}{C_1}$

It is also worth to point out that the offset suppression factor for all AZ and CDS circuits is related to $1/A$, while for UC circuits (a) and (b), they are all greater than 1. Therefore, the DC offset of opamp by AZ and CDS will be well eliminated.

For corroborating the effectiveness of these derived expressions and results, both the ASIZ [4.30] and SWITCAP [4.31] simulation outcomes with respect to both gain and phase errors have been computed in Figure 4-4(a) and (b) respectively, where $A=100$ and all related capacitance ratios are unity. The simulated results show excellent agreement with the theoretical expressions: the employment of the MF configuration results in half of the gain error of the conventional UC SC circuit; and all CDS circuits exhibit a different-level compensation in finite gain errors. The H-CDS circuits have only effective gain compensation when signal frequency is lower than a quarter of the sampling rate when compared to the UC MF circuit. On the other hand, SC/H-CDS circuit of Figure 4-3(d) is only comparable to EC/H-CDS of Figure 4-3(e) when k_1 is fairly small (=0.1). The EC/P-CDS circuits of Figure 4-3(g), (h) and (j) achieve the highest compensation precision for their smallest, frequency-independent gain and phase errors. When k_1 and k'_h are unity, SC/P-CDS circuit of Figure 4-3(f) and EC/P-CDS of Figure 4-3(i) operate at the similar level of EC/H-CDS of Figure 4-3(e) in performance, only when these ratios become smaller, e.g. 0.1, their performance will have now 10 times improvement, which is consistence with the theoretical expectation.

Note that all circuits are no longer stray-insensitive due to the finite gain, which, of course, will relatively degrade the compensation performance. Moreover, employing error-storage capacitor in CDS circuits will slightly increase the parasitic effect due to an extra parasitic at the super virtual ground node x. The simulated gain and phase errors of all the above circuits are further illustrated in Figure 4-4(c) and (d) where the parasitics with 10% and 30% of the capacitance for the top and bottom plate of each capacitor, and with also a non-zero input capacitor of opamp. This input capacitor becomes not negligible due to the enlarged g_m normally needed for high-frequency circuits. The results clearly show the increase in the magnitude and phase errors of all the above circuits, especially the circuit in Figure 4-3(e) and (h). The significant degradation for circuit in Figure 4-3(h) is mainly because that the parasitics at node x will draw twice the charges from the sampling capacitor C_1 due to the two-step P-CDS process. The simulation shows that if the top-plate of C_h is connected to node x instead of virtual ground, it will help a little bit the elimination of this effects in this circuit.

In addition, although there is no capacitor mismatches contribution to the delay unity-gain of the circuit in Figure 4-3(j), however, the output voltage in the calibration phase, where the error voltage is stored, depends on the capacitor ratio accuracy, i.e. $k_2 = C_2/C_F$. This effect, nevertheless, is fairly small especially when compared to H-CDS, and the simulation is also performed as shown in curve j' for the ratio having 3% variation, i.e. $k_2 = 0.97$ where shows that the average error is increased although the dependency with frequency is slightly reduced, because the magnitude factor of cosine shape function is also proportional to k_2 as shown in the Table 4-1.

3.3 Multi-Unit Delay Implementations

As discussed in the last chapter polyphase structures can be optimized by adjusting the achieved delay of ADB, thus the flexible implementation of multi-unit delay for delay blocks is necessary for an optimum design. This session will propose several different realizations of SC delay circuits with only one opamp but flexible delay arrangements based on the CDS techniques presented in the last session and also the parallel rotating-switching techniques.

Both AZ and SC/H-CDS circuits techniques cannot provide compensated output at the input sampling phase, therefore, only a rational number of delays can be achieved, e.g. 1.5, 2.5, etc. Presented in Figure 4-5(a) and (b) are the AZ and SC/H-CDS MF SC circuit realizations with 1.5 unit delays by rotating switching two parallel SC sampling branches.

Figure 4-5(c) and (d) are proposed to implement EC/H-CDS and SC/P-CDS MF SC delay circuits with one unity delay. Especially, due to the usage of the extra error-storage capacitor, the output phase can be assigned in a more flexible way, like the circuit of Figure 4-5(c) can also realize 1.5 unit delay with parenthesized phases. It can be straightforwardly extended to using n input branches for realizing delay both z^{-n} and $z^{-(n+1/2)}$. Note that the original two prediction branches are efficiently combined into one for the circuit of Figure 4-5(d), and in addition, in this special configuration, no S/H input is required for this circuit.

As mentioned before, SSC can be easily applied in conventional SC circuit with just an extra SSC branch and two more phases, Figure 4-5(e) and (f) are the proposed examples of the double-unit SC delay circuits in UC and EC/P-CDS configurations.

All the above structures can be further extended to realize more unit delays with more parallel branches and clock phases, but with efficiently only one opamp.

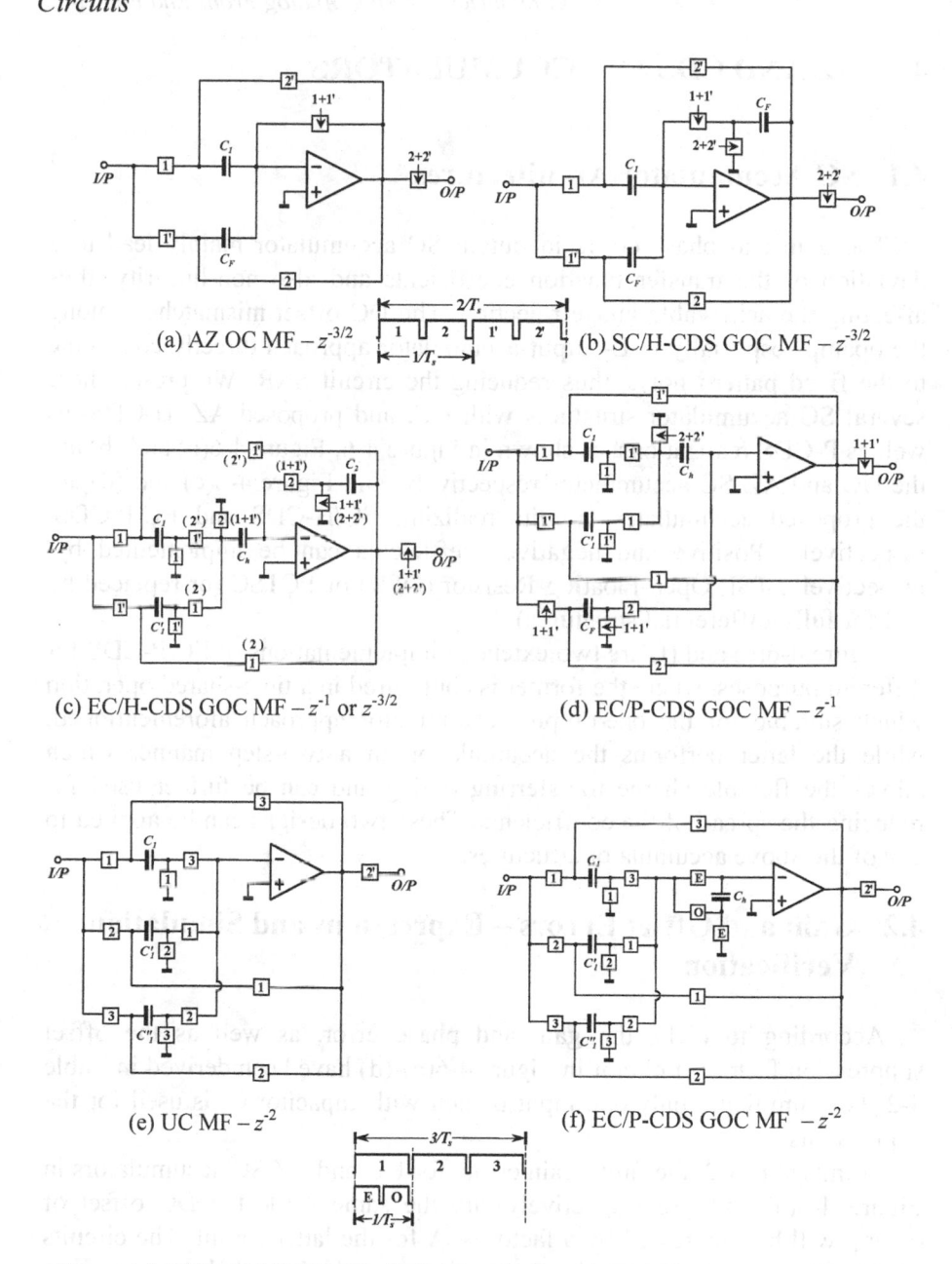

(a) AZ OC MF $- z^{-3/2}$

(b) SC/H-CDS GOC MF $- z^{-3/2}$

(c) EC/H-CDS GOC MF $- z^{-1}$ or $z^{-3/2}$

(d) EC/P-CDS GOC MF $- z^{-1}$

(e) UC MF $- z^{-2}$

(f) EC/P-CDS GOC MF $- z^{-2}$

Figure 4-5. Different MF UC, AZ, CDS delay blocks with flexible delay implementation

4.   AZ AND CDS SC ACCUMULATORS

4.1   SC Accumulator Architectures

The gain and phase errors in output SC accumulator mainly lead to a deviation of the transfer function coefficients and also non-linearity, thus affecting the achievable image rejection. The DC offset mismatches among the opamps especially in L-output-accumulator approach directly contribute to the fixed pattern noise, thus reducing the circuit SNR. We present here several SC accumulator structures with UC, and proposed AZ, H-CDS, as well as P-CDS realizations, as shown in Figure 4-6. Figure 4-6(a) and (b) are the UC and AZ SC accumulator respectively, and Figure 4-6(c) and (d) are the proposed accumulator circuits realizing EC/H-CDS and EC/P-CDS, respectively. Positive and negative coefficients can be implemented by, respectively, TSI, Open-Floating Resistor (OFR) or PCTSC (or replaced by TSI for fully-differential structures).

Figure 4-6(e) and (f) are two extended implementations of EC/P-CDS for different purposes, where the former is configured in a time-shared operation which suitable for the one-output accu-mulator approach aforementioned, while the latter performs the accumulation in a two-step manner which allows the flexible charge transferring timing and can be further used for reducing the spread of the coefficients. These two designs can be applied to any of the above accumulator structures.

4.2   Gain and Offset Errors – Expressions and Simulation Verification

According to (4.1), the gain and phase error, as well as the offset suppression factor for circuit in Figure 4-6(a)-(d) have been derived in Table 4-2. For simplicity, only one input branch with capacitor C_1 is used for the expressions.

From Table 4-2, the finite-gain errors for UC and AZ SC accumulators in Figure 4-6(a) and (b), respectively, are the same while the DC offset of opamp will be suppressed by a factor of A for the latter circuit. The circuits EC/H-CDS and EC/P-CDS in Figure 4-6(c) and (d) both have less gain errors and also 1/A suppression factor for the DC offset. Like those for the delay circuits, their gain and phase errors are dependent of frequency and described respectively by cosine and sine function with different weighted magnitude. Especially, this weighted factor for the EC/H-CDS circuit depends on real summing gain factor of the accumulator or the implemented coefficient, while that for EC/P-CDS is independent of the gain of the summing circuit but instead depends on the ratio of error-storage capacitor and the summing capacitor.

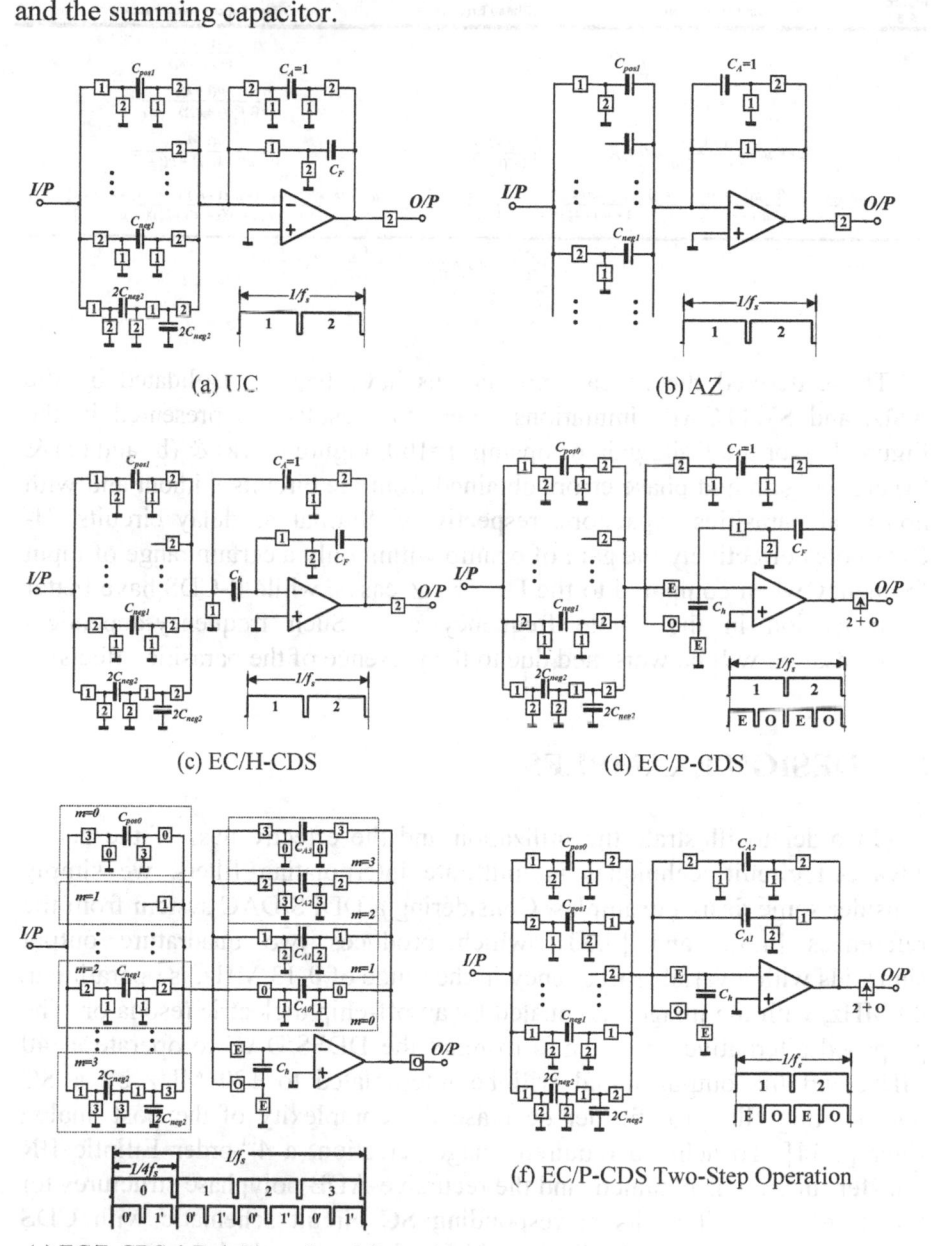

(a) UC

(b) AZ

(c) EC/H-CDS

(d) EC/P-CDS

(e) EC/P-CDS 4-Path Time-shared Operation

(f) EC/P-CDS Two-Step Operation

Figure 4-6. Different SC accumulator architectures with UC, AZ and CDS techniques

Table 4-2. Gain & phase errors and offset-suppression factor for SC accumulator circuits in Figure 4-6 (a)-(d)

Figure 4-6	Gain Error $m(j\omega)$	Phase Error $\theta(j\omega)$	Offset Suppression Factor γ
(a)	$-(1+k_1)\mu$	0	$\dfrac{(1+k_1)}{1+\mu(1+k_1)}$
(b)	$-(1+k_1)\mu$	0	$\dfrac{\mu(1+k_1)}{(1+\mu(1+k_1))(1+\mu)}$
(c)	$-(1+k_1)\mu+\dfrac{(1+k_1)\mu}{1+(1+k'_h)\mu}\cos(\omega T_s)$	$-\dfrac{(1+k_1)\mu}{1+(1+k'_h)\mu}\sin(\omega T_s)$	$\dfrac{\mu(1+k_1)}{(1+\mu(1+k_1))(1+\mu(1+k'_h))}$
(d)	$-\dfrac{\mu((1+k_1+k_h)(1+\mu(1+k_1))-(1+k_1))}{1+2\mu(1+k_1)}+\dfrac{\mu k_h\cos(\omega T_s)}{1+\mu(1+k'_h)}$	$-\dfrac{\mu k_h\sin(\omega T_s)}{1+\mu(1+k'_h)}$	$\dfrac{\mu(1+2k_1+k_1^2-k_1k_h)+\mu^2(1+k_1)(1+k_1+k_h+k'_h+k_1k'_h-k_hk'_h)}{(1+\mu(1+k_1+k_h))(1+\mu(1+k'_h))(1+\mu(1+k_1))}$

$$\text{where}\quad \mu=\frac{1}{A},\quad k_1=\frac{C_1}{C_A},\quad k_A=\frac{C_A}{C_F},\quad k_h=\frac{C_h}{C_A},\quad k'_h=\frac{C_h}{C_F}$$

Those derived theoretical expectations have been consolidated by the ASIZ and SWITCAP simulations where the results are presented in the Figure 4-7 for the finite gain of opamp A=100. Figure 4-7(a) & (b) and (c) & (d) are the gain and phase errors obtained from the circuits without and with non-ideal parasitics capacitors, respectively. Similar as delay circuits, H-CDS boost effectively the gain of opamp within only a certain range of input frequency when compared to the UC or AZ cases, while P-CDS have better compensation in the whole frequency axis. Such frequency-dependent compensation will be worsened due to the presence of the parasitic effects.

5.   DESIGN EXAMPLES

In order to illustrate the utilization and the effectiveness of the above advanced circuit techniques in multirate interpolating filters, we simply consider some design examples. Considering a DDFS/DAC system from the references [4.32] and [4.33], which produces two quadrature output sinusoids with a variable frequency in the range of 0-13 MHz, is operating at 80 MHz, with the images attenuated by an off-chip dielectric resonator. The proposed alternative approach is to relax the DDFS/DAC to operate at 40 MHz, and the output signal will be interpolated to 120 MHz by a SC interpolating filter for further decrease the complexity of the post analog filter [4.34]. To achieve required image rejection, a 4th-order Elliptic IIR transfer function is obtained, and the recursive-ADB polyphase structures for this filter, as well as its corresponding SC circuit schematic with CDS techniques are presented in Figure 4-8(a) and (b) respectively.

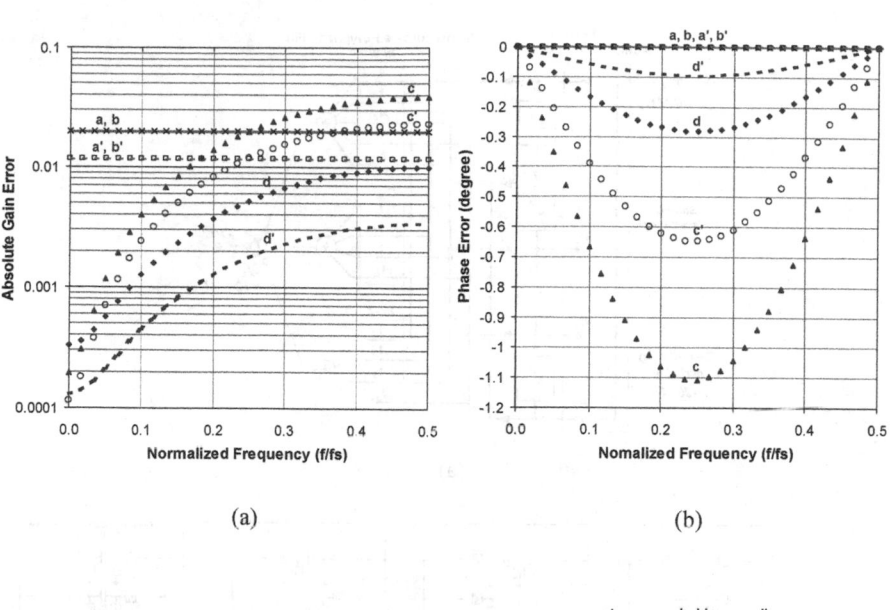

(a)   (b)

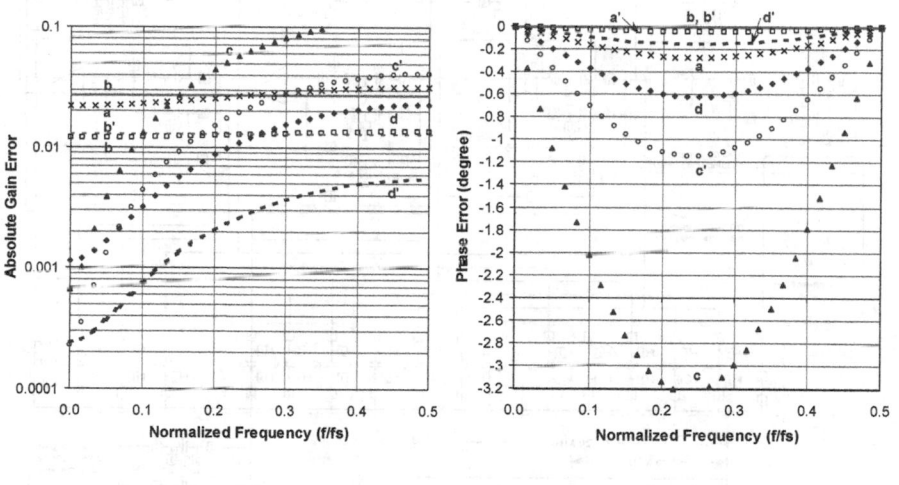

(c)   (d)

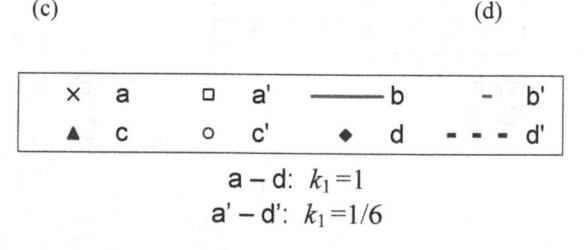

a – d: $k_1 = 1$

a' – d': $k_1 = 1/6$

Figure 4-7. Simulated gain & phase errors for SC accumulator circuits in Figure 4-6 without parasitics (a) & (b) and with parasitics (c) & (d)

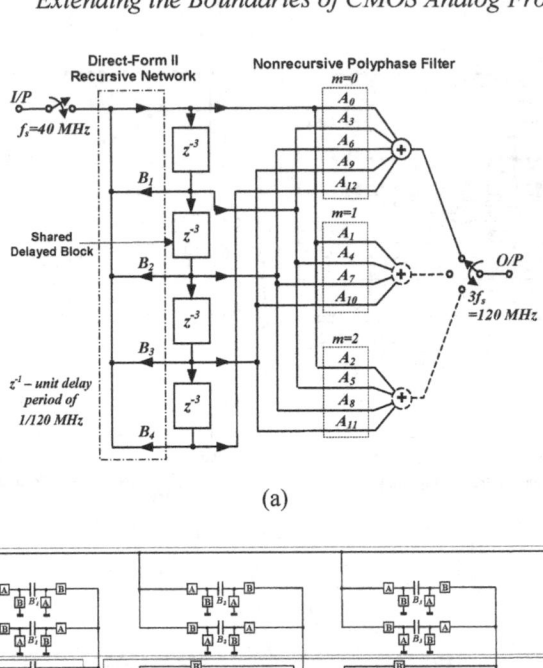

(a)

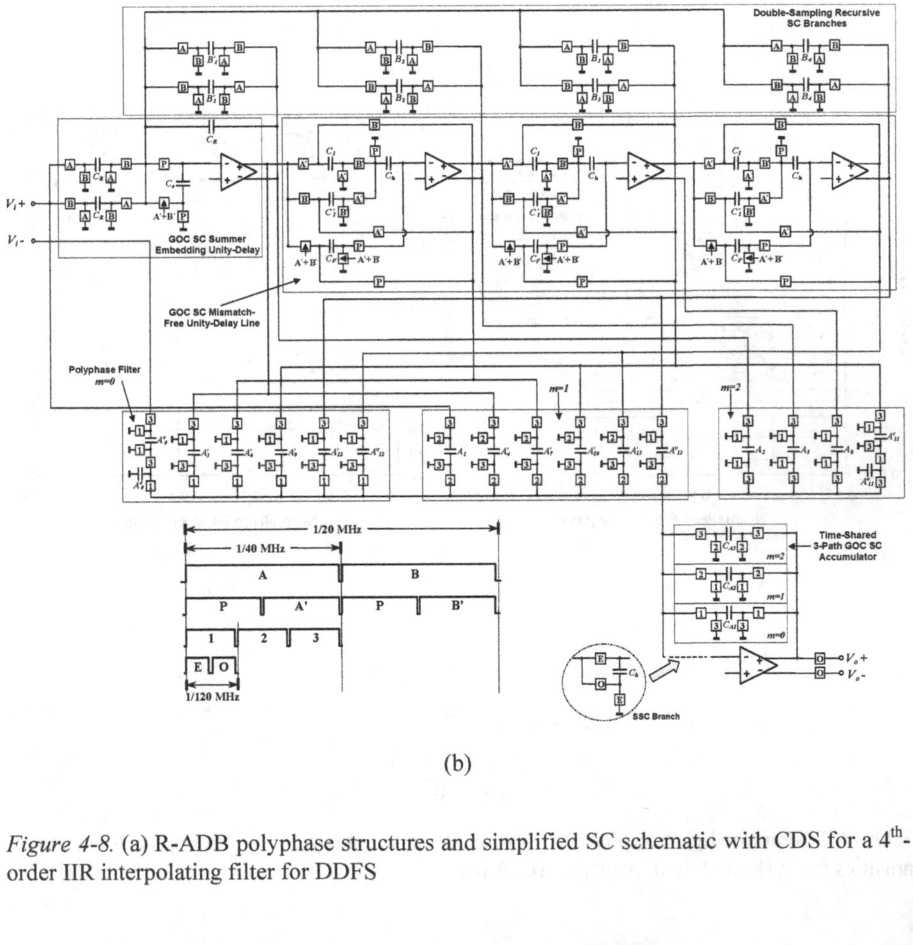

(b)

Figure 4-8. (a) R-ADB polyphase structures and simplified SC schematic with CDS for a 4th-order IIR interpolating filter for DDFS

The accuracy of the serial ADB delay line here is very critical stage due to the fact that it will be shared for both numerator and denominator polynomials of the IIR transfer function, which affects the locations of both poles and zeros, and most importantly, degrade the pole-zero cancellation owing to the multirate transformation [4.35, 4.36, 4.37] producing sharp undulations in the stopband. Consequently, the P-CDS techniques with mismatch-free delay blocks, as presented in Figure 4-5(d), are employed in this ADB serial delay line for minimizing the accumulated errors. Among all the ADB stages, the first delay block, which additionally serves as summer as well for adding the signals from recursive branches, is the most error-sensitive one. Thus this multifunction SC summer/delay circuit is realized by using SSC summing property, as presented in Figure 4-6(d) for achieving high-accuracy P-CDS. In order to alleviate the increased speed requirements of opamp due to the CDS nature, double-sampling techniques are also efficiently manipulated in recursive networks. It will only efficiently double in a small amount SC branches which number is equal to the filter order, i.e. 4, for our design, thus maintaining the same settling time requirement of half period of lower input sampling period 12.5 ns.

One-time-shared output accumulator with three multiplexed summing paths can be designed for all polyphase filters, since mismatches in both the gain and offset errors among the polyphase subfilter paths is rather much more important than the absolute error, UC circuit can be simply employed without too much illness, as it will be discussed later. Moreover, it is also worth to point out that it can achieve the P-CDS gain and offset compensation by just simply adding a SSC branch to the traditional UC accumulator structure but with the trade-off of two additional clock phases E and O as well as reduced time slots for settling.

The simulations of this circuit are illustrated in the Figure 4-9 with also the comparison to the ideal and uncompensated one. The simulations are performed with opamp gain of 300, and especially a input parasitics of 0.5 pF presented at the opamp input. It is clearly evident that the variation of the pole and zero placement due to the finite gain error and parasitics for the circuit using EC/P-CDS is very small, and thus amplitude response is almost fully matched to the ideal one, and even if for the case that there is no compensation techniques used in the last output accumulator, the passband and stopband deviation (only little net gain shift) is still satisfactory and closed to the one with EC/P-CDS. All the unwanted image bands located at 27-40 MHz, 40-53 MHz, 67-80 MHz and 80-93 MHz have been attenuated about 60 dB. Especially, when comparing to the uncompensated circuit with high gain of 5000, the proposed circuit has the equivalent passband but with much smaller stopband deviation. While for the uncompensated one with gain of 300, its passband and especially stopband varies more than 1.6 dB and 20 dB respectively due to the large distortion of the placement of poles and zeros.

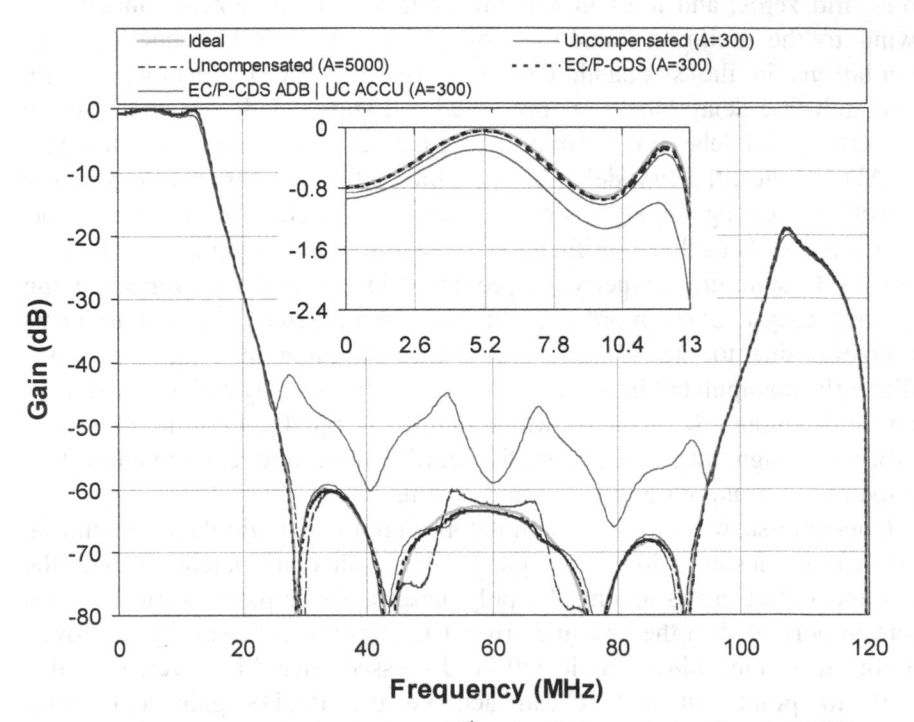

Figure 4-9. Simulated amplitude response of 4th-order IIR interpolating filter for DDFS

Another example is an FIR interpolating filter which raises the sampling rate by 4 fold and achieves the 15-tap lowpass filtering with the passband ripple less than 0.2dB and image attenuation greater than 50dB. The simulated results of the circuits employing the H-CDS and P-CDS unity-delay ADB and accumulator structures discussed in previous sessions are presented in Figure 4-10 with opamp gain of 100 only. It is clearly evident that the transmission zeros variation of P-CDS circuit due to the finite gain error is extremely small as plotted in the zeros patterns of Figure 4-10(a) and thus the amplitude response is almost fully matched to the ideal one as shown in Figure 4-10(b). While for the uncompensated one, its maximum passband and stopband variation are more than 0.7dB and 10dB respectively due to the seriously distorted location of zeros shown in Figure 4-10(a). Besides, the response of the interpolator with H-CDS is better than the uncompensated but worse than P-CDS circuit due to its inherent narrow-band compensation property.

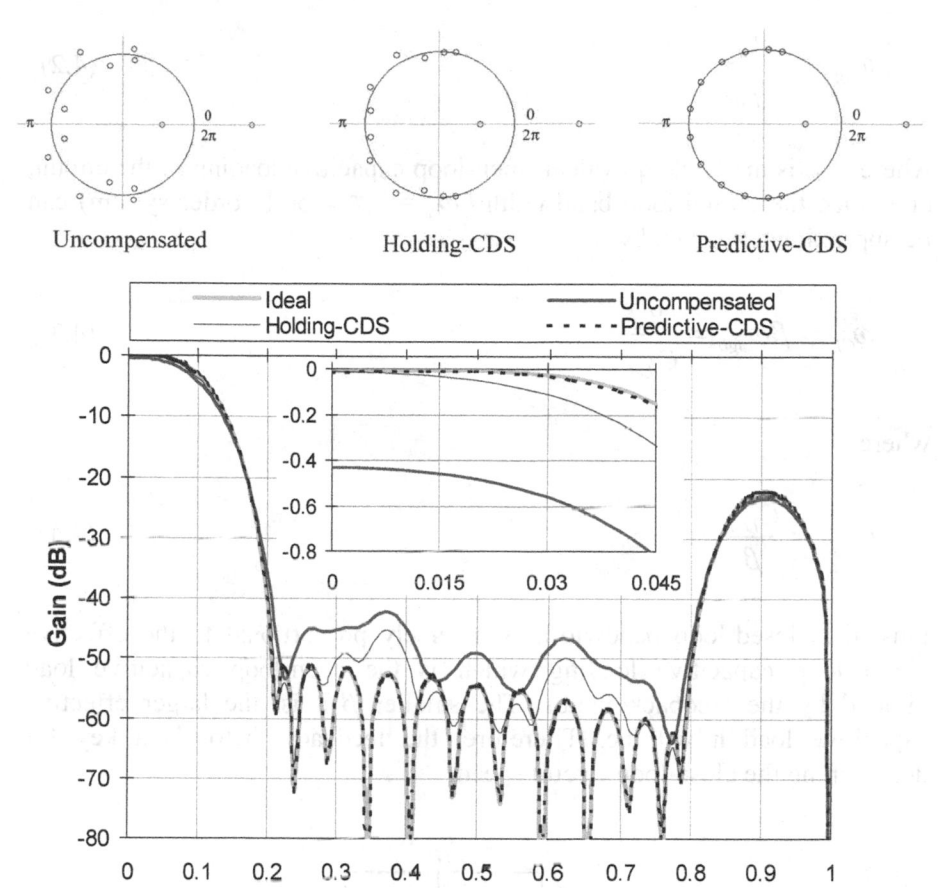

Figure 4-10. (a) Zero plots and (b) Simulated amplitude response of a 15-tap SC FIR interpolating filter with UC, H-CDS and P-CDS realizations (A=100)

6.   SPEED AND POWER CONSIDERATIONS

As analyzed in Chapter 3 / Session 2, the speed and power of SC circuits is directly proportional to the equivalent total capacitive loading. Since the AZ and CDS circuits possesses more circuit elements and clock phases than the conventional UC one, the speed and power requirements will be different. This session presents the speed and power analysis for AZ and CDS circuits.

Assuming that the single-stage one-pole transconductance opamps, which are widely used in modern SC circuits, are employed, the unity-gain bandwidth can be expressed by

$$\omega_{ugb} = \frac{g_m}{C_{Ltot}} \qquad (4.2)$$

where C_{Ltot} is the total equivalent open-loop capacitive loading of the circuit, and since the closed-loop bandwidth ($\omega_{CL} = 1/\tau_{CL}$ for 1st-order system) can be approximately given by

$$\omega_{CL} = \beta\omega_{ugb} = \frac{g_m}{C_{Leff}} \qquad (4.3)$$

where

$$C_{Leff} = \frac{C_{Ltot}}{\beta} \qquad (4.4)$$

thus, the closed-loop bandwidth is inversely proportional to the effective closed-loop capacitive loading which is the open-loop capacitive load divided by the feedback factor. The smaller β it is, the larger effective capacitive load it will be. Therefore, the feedback factor is a key for determining the close-loop circuit speed.

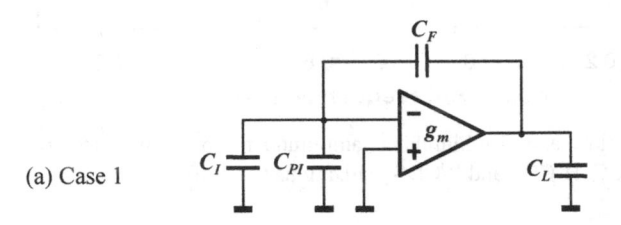

(a) Case 1

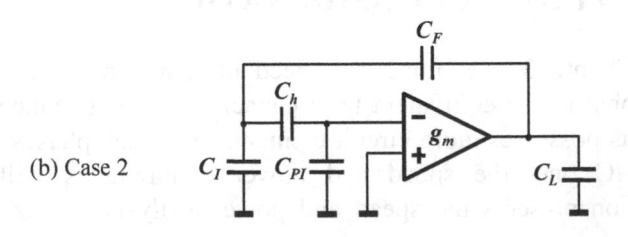

(b) Case 2

Figure 4-11. Circuit configurations for different operation phases for UC, AZ and CDS SC circuits

Figure 4-11(a) and (b) generally describe circuit configurations for UC, AZ and CDS circuits in different operation phases where C_I and C_{PI} are the total capacitance in input branches (also including the parasitics at this node) and total parasitic capacitance at virtual ground including the input capacitance of opamp, respectively. On the other hand, C_F and C_L are the feedback capacitor and total output capacitive loading including also the output parasitic capacitance of opamp. Especially, C_h is the error storage capacitor. The former Figure 4-11(a) can be used for characterizing all UC, AZ and SC/CDS and the calibration/error storage phase for EC/CDS SC delay and accumulator circuits, while Figure 4-11(b) describes all the EC/CDS circuits during the output/operation phase. For instance, in phase 1 of AZ delay circuit in Figure 4-3(c), corresponding to the equivalent circuit in Figure 4-11(a), C_I represents C_1 and its top-plate parasitic capacitance, C_F becomes infinity, while during phase 2, $C_F=C_1$ and C_I can be the bottom-plate parasitic capacitance of C_1. Furthermore, in phase 1 of EC/P-CDS circuit of Figure 4-3(j), corresponding also to the equivalent circuit in Figure 4-11(a): C_I represents C_2+C_h and their top-plate parasitic capacitance, while in phase 2, corresponding to the circuit in Figure 4-11(b). Here, C_I represents the top- and bottom-plate parasitic capacitance of C_1 and C_h, respectively, while $C_F=C_1$, C_{PI} is the input capacitance of the opamp and the top-plate parasitic capacitance of C_h. This can be similarly applied to any other delay or accumulator circuits.

According to Figure 4-11(a) and (b), their corresponding open-loop total capacitive loading and feedback factor can be obtained respectively by

$$C_{Ltot,1} = C_L + \beta_1 \cdot (C_I + C_{PI}) \tag{4.5}$$

$$\beta_1 = \frac{C_F}{C_I + C_{PI} + C_F} \tag{4.6}$$

and

$$C_{Ltot,2} = C_L + \beta_2 \cdot \left(C_I(1 + \frac{C_{PI}}{C_h}) + C_{PI} \right) \tag{4.7}$$

$$\beta_2 = \frac{C_F}{C_I + C_{PI} + C_F + \frac{C_{PI}}{C_h}(C_I + C_F)} \tag{4.8}$$

Comparing (4.6) and (4.8), we can deduce that only if there is no input capacitance of opamp, the two cases will be exactly the same. However, the presence of this capacitance leads to an increase of the feedback factor in the case 2 for the CDS circuits with the employment of error-storage capacitor C_h, thus equivalently rendering the increase of the close-loop effective capacitive loading. In other words, for a required closed-loop time constant, this increase directly results in a higher value necessary for the transconductance g_m, which is also directly proportional to the required power dissipation according to (3.3). Therefore, the ratio C_{PI}/C_h should be minimized.

In order to clarify the speed and power effects due to usage of the error-storage capacitor, Figure 4-12 presents the comparison in terms of the feedback factor β and C_{Leff} for a typical example with $C_I=C_F=C_{PI}=1$ pF and $C_L=2$ pF vs. the ratio C_{PI}/C_h. The results show that if C_{PI} is greater than C_h 10 times, the $C_{Leff,2}$ is nearly as large as 60 pF and the β_2 is also almost 10 times smaller than β_1, thus the required g_{m2} is also more than 7 times larger than g_{m1}, even if $C_{PI}/C_h=1$, the β_2 and $C_{Leff,2}$ still has about 0.6 and 1.6 times to those in the case 1.

According to the (3.2) and (3.3), Figure 4-12(a) presents the comparison to the required tail current for case 1 and case 2 assuming that the telescopic opamp is used. The case 2 needs more than 1.6 and 7 times the amount of the current of case 1 for achieving the enlarged g_m when $C_{PI}/C_h=1$ and 10, respectively. Note that the current from SR requirement for case 2 is slightly smaller (maximum 5 %) than that for case 1. Therefore, for EC/CDS circuits, only when the circuit has very high SR requirement (meaning that the power of the system could be dominated from this SR requirement with also a sufficiently small C_{PI}/C_h ratio) then the case 1 and case 2 can be comparable. For example, assuming 1/5 of the phase duration to be allocated for slewing while the remaining 4/5 for linear settling and $V_{eff}=200$ mV, it can be easily obtained that only when V_{ostep} is greater than $0.35/\beta$, then the tail current will be dominated by that required from SR and both case 1 and case 2 require similar power consumption, e.g. if $V_{ostep}=1$ V, it needs C_h greater than C_{PI} 10 times to ensure feedback factor at 0.35 from curve in the Figure 4.12(a). Note that this is not practical for real implementation, due to the fact that such large C_h will first increase the power consumption during the calibration phase where C_h functions as a parasitic capacitor at virtual ground of opamp, and secondly it will also reduce the accuracy of the gain compensation as the C_h/C_F is enlarged at the same time.

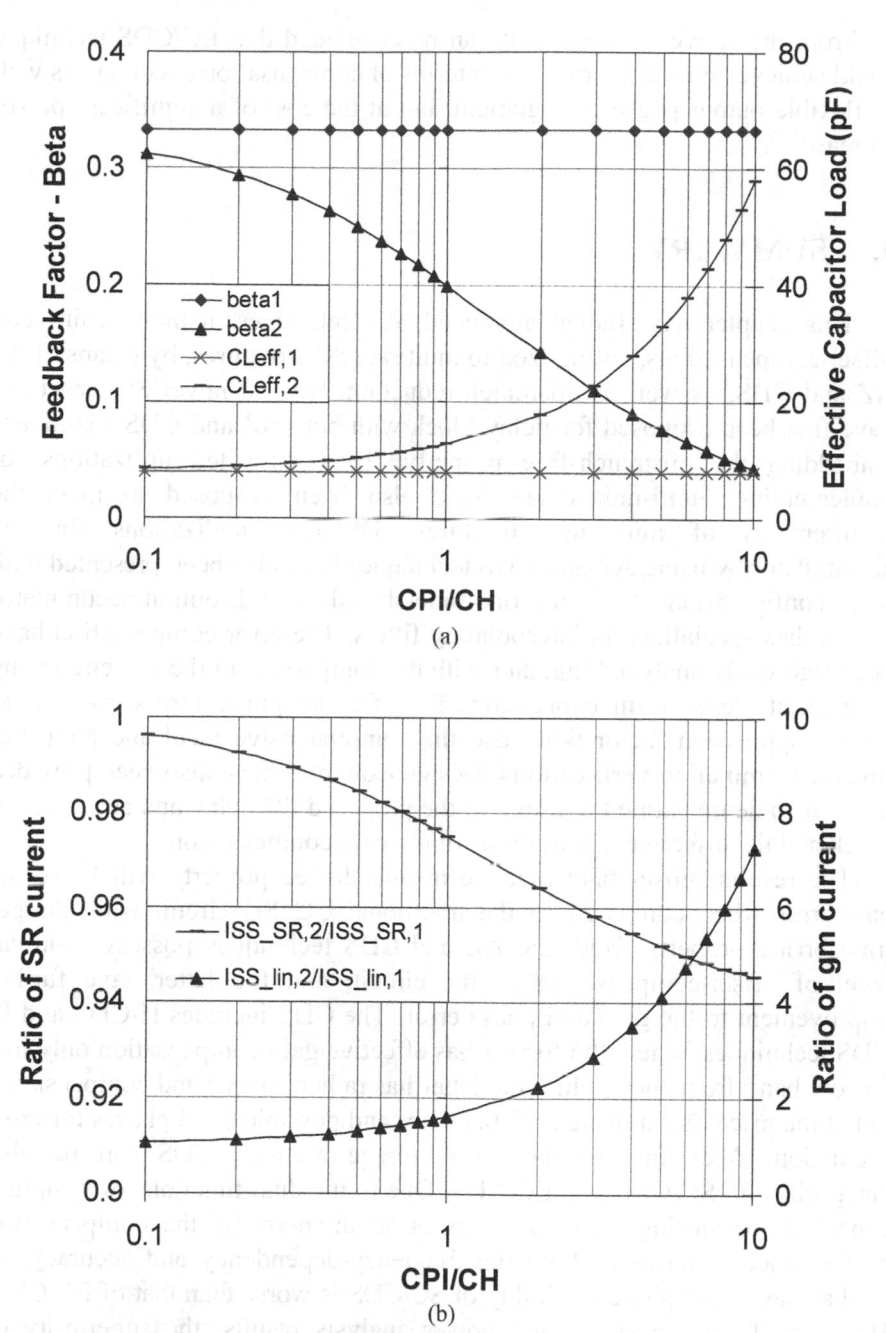

Figure 4-12. (a) Feedback factor and effective capacitive loading (b) Current consumption for SR and linear settling versus C_{PI}/C_h for CDS circuits with employment of error-storage capacitor

From the above discussion, it can be concluded that EC/CDS technique could achieve better performance in terms of compensation accuracy, as well as flexible output phase arrangement, but at the cost of a significant power increase.

7.   SUMMARY

This chapter has studied advanced SC techniques, namely, gain- and offset-compensations, specialized to multirate SC structures, by means of the AZ and CDS, as well as mismatch-reduction. Several novel SC structures have first been proposed for delay block with both AZ and CDS techniques embedding the mismatch-free property. Their extended utilizations for implementing multi-unit delay have also been proposed to meet the requirements of multirate structures. Different realizations for SC accumulator by using AZ and CDS techniques have also been presented with their configurations for both one-time-shared- and L-output-accumulator approaches specialized in interpolating filters. The error compensation have been rigorously analyzed together with the comparison to the UC circuits by both exact closed-form expressions for gain and phase errors, as well as offset suppression factor with also the comprehensive ideal and parasitic-involved simulation verifications. Design examples have also been provided in order to demonstrate the usage of the proposed SC delay and accumulator block and the effectiveness in terms of the gain compensation.

The results substantiate that the mismatch-free property will halve the gain-error when compared to the traditional UC SC circuit with charge-transferring property. Both the AZ and CDS techniques possess a similar level of offset-compensation to the circuit, and the latter have further improvement to the gain and phase error. The CDS includes H-CDS and P-CDS techniques, where the former has effective gain compensation only in a narrow band frequency, while the latter has rather wider band compensation but at the price of duplicated SC branches and possible extra phases for error prediction. According to the error storage method, CDS can be also categorized in SC/CDS and EC/CDS. Due to the dual functions of sampling capacitor for storing signal and errors simultaneously, the compensation performance in terms of both the frequency-dependency and accuracy, as well as the output phase flexibility of SC/CDS is worse than that of EC/CDS. However, from the speed and power analysis results, the superiority of EC/CDS is built on the price of the significantly increased speed and power requirements due to the increased effective total capacitance loading imposed by the existing parasitic capacitance in practice, especially at the opamp input. Therefore, it finally can be suggested that the EC/CDS techniques are not very suitable for high-frequency applications due to their high power consumption nature, and the AZ and SC/CDS will be the best choice for high-frequency applications especially those not so sensitive to, either the finite gain or to a very narrow signal band.

REFERENCES

[4.1] Christian C.Enz, G.C.Temes, "Circuit techniques for reducing the effects of Op-Amp imperfections: Autozeroing, Correlated Double Sampling, and Chopper Stabilization," *Proceeding of The IEEE*, Vol.84, No.11, pp.1584-1614, Nov.1996.

[4.2] R.C.Yen, P.R.Gray, "A MOS Switched-Capacitor instrumentation amplifier," *IEEE J. Solid-State Circuits*, Vol.SC-17, pp.1008-1013, 1982.

[4.3] Y.Huang, P.Ferguson, G.C.Temes, "Reduced Nonlinear Distortion in Circuits with Correlated Double Sampling," *IEEE Trans. Circuits and Systems – II: Analog and Digital Signal Processing*, Vol.44, No.7, pp.593-597, Jul.1997.

[4.4] K.K.K.Lam, M.A.Copeland, "Noise-cancelling Switched-Capacitor filtering technique," *IEE Electronics Letters*, Vol.19, No.20, pp.810-811, Sep.1983.

[4.5] K.Haug, F.Maloberti, G.C.Temes, "Switched-Capacitor integrators with low finite-gain sensitivity," *IEE Electronics Letters*, Vol.21, No.24, pp.1156-1157, Nov.1985.

[4.6] K.Haug, F.Maloberti, G.C.Temes, "Switched-Capacitor circuits with low op-amp gain sensitivity," *Proc. IEEE International Symposium on Circuits and Systems (ISCAS)*, No.24, pp.797-800, May.1986.

[4.7] K.Nagaraj, K.Singhal, T.R.Viswanathan, J.Vlach, "Reduction of finite-gain effect in Switched-Capacitor filters," *IEE Electronics Letters*, Vol.21, No.15, pp.644-645, Jul.1985.

[4.8] Wing-Hung Ki, G.C.Temes, "Offset-Compensated Switched-Capacitor Integrator," in *Proc. IEEE International Symposium on Circuits and Systems (ISCAS)*, pp.2829-31, 1990.

[4.9] Wing-Hung Ki, G.C.Temes, "Gain- and Offset-compensated Switched-Capacitor filters," in *Proc. IEEE International Symposium on Circuits and Systems (ISCAS)*, pp.1561-15664, 1991.

[4.10] L.E.Larson, K.W.Martin, G.C.Temes, "GaAs Switched-Capacitor circuits for high-speed signal processing," *IEEE J. Solid-State Circuits*, Vol.SC-22, No.6, pp.971-980, Dec.1987.

[4.11] A.Nagari, A.Baschirotto, F.Montecchi, R.Castello, "A 10.7-MHz Bi-CMOS high-Q double-sampled SC bandpass filter," *IEEE J. Solid-State Circuits*, Vol.SC-32, No.10, pp.1491-1498, Oct.1997.

[4.12] L.E.Larson, G.C.Temes, "Switched-Capacitor gain stage with reduced sensitivity to finite amplifier gain and offset voltage," *IEE Electronics Letters*, Vol.22, pp.1281-82, Nov.1985.

[4.13] K.Nagaraj, T.R.Viswanathan, K.Singhal, J.Vlach, "Switched-Capacitor circuits with reduced sensitivity to amplifier gain," *IEEE Trans. on Circuits and Systems*, Vol.CAS-34, pp.571-574, May 1987.

[4.14] H.Shafeeu, A.K.Betts, J.T.Taylor, "Novel amplifier gain insensitive Switched Capacitor integrator with same sample correction properties," *IEE Electronics Letters*, Vol.27, No.24, pp.2277-2279, Nov.1991.

[4.15] G.Palmisano, G.Palumbo, "High-frequency differential SC integrator with highly accurate gain compensation," *International J. of Circuit Theory and Applications*, Vol.22, pp.71-75, 1994.

[4.16] G.C.Temes, Huang Yunteng, P.F. Ferguson, "A high-frequency track-and-hold stage with offset and gain compensation," *IEEE Trans. on Circuits and Systems-II*, Vol.42, pp.559-561, Aug 1995.

[4.17] J.A.Grilo, G.C.Temes, "Predictive correlated double sampling Switched-capacitor integrators," in *Proc. International Conference on Electronics, Circuits and Systems (ICECS)*, vol.2, pp.9-12, Sep. 1998.

[4.18] Huang Yunteng, G.C.Temes, P.F. Ferguson, "Offset- and gain-compensated track-and-hold stages," in *Proc. International Conference on Electronics, Circuits and Systems (ICECS)*, vol.2, pp.13-16, Sep. 1998.

[4.19] Xiaojing Shi, H.Matsumoto, K.Murao, "Gain- and offset-compensated non-inverting SC circuits," in *Proc. IEEE International Symposium on Circuits and Systems (ISCAS)*, vol.2, pp.425 -428, 2000.

[4.20] P.Gillingham, "Stray-free Switched-Capacitor unit-delay circuit," *IEE Electronics Letters*, Vol.20, No.7, pp.308-310, Mar 1984.

[4.21] K.Nagaraj, "Switched-Capacitor delay circuit that is insensitive to capacitor mismatch and stray capacitance," *IEE Electronics Letters*, Vol.20, No.16, pp.663-664, 1984.

[4.22] A.E.Said, "Stray-free Switched-Capacitor building block that realizes delay, constant multiplier, or summer Circuit," *IEE Electronics Letters*, Vol.21, No.4, pp.167-168, Feb.1985.

[4.23] A.Dabrowski, U.Menzi, G.S.Moschytz, "Offset-compensated Switched-Capacitor delay circuit that is insensitive to stray capacitance and to capacitor mismatch," *IEE Electronics Letters*, Vol.25, No.10, pp.623-625, May 1989.

[4.24] J.J.F.Rijns,.H.Wallinga, "Stray-insensitive switched-capacitor sample-delay-hold buffers for video frequency applications," *IEE Electronics Letters*, Vol.27 Issue: 8 , pp.638-640, Apr.1991.

[4.25] S.Eriksson, "Realization of Switched-Capacitor delay lines and Hilbert transformers," *IEE Electronics Letters*, Vol.27, No.14, pp.1262-1264, Jul.1991.

[4.26] Hiroshi Iwakura, "Realization of Tapped Delay Lines Using Switched-Capacitor LDI Ladders and Application to FIR Filter Design," *IEEE Trans. on Circuits and Systems-II*, Vol.40, pp.794-797, Dec. 1993.

[4.27] Seng-Pan U, R.P.Martins, J.E.Franca, "Highly accurate mismatch-free SC delay Circuits with reduced finite gain and offset sensitivity," in *Proc. IEEE International Symposium on Circuits and Systems (ISCAS)*, Vol.2, pp.57-60, USA, May 1999.

[4.28] Seng-Pan U, R.P.Martins, J.E.Franca, "Offset- and gain compensated and mismatch-free SC delay circuit with flexible implementation," *IEE Electronics Letters*, Vol.35, No.3, pp.188-189, Feb. 1999.

[4.29] K.Martin, A.S.Sedra, "Effects of the Op-Amp finite gain and bandwidth on the performance of Switched-Capacitor filters," *IEEE Trans. on Circuits and Systems*, Vol.CAS-28, pp.822-829, Aug.1981.

[4.30] A.C.M.Queiroz, P.M.Pinheiro, L.P.Calôba, "Systematic nodal analysis of switched-current filters," *Proc. IEEE International Symposium on Circuits and Systems (ISCAS)*, pp. 1801-1804, Jun. 1991.

[4.31] K.Suyama, S.C.Fang, "User's Manual of SWITCAP2 Version 1.1," Columbia University, 1992.

[4.32] G.Chang, A.Rofougaran, Mong-Kai Ku, A.A.Abidi, H.Samueli, "A Low-Power CMOS Digitally Synthesized 0-13MHz Agile Sinewave Generator," in *IEEE ISSCC Digest of Technical Paper*, pp.32-33. Feb. 1994.

[4.33] A.Rofougaran, G.Chang, J.J.Rael, J.Y.-C.Chang, M. Rofougaran, P.J.Chang, M.Djafari, M-K Ku, E.W.Roth, A.A.Abidi, H.Samueli,"A Single-Chip 900-MHz Spread-Spectrum Wireless Transceiver in 1-μm CMOS- Part I: Architecture and Transmitter Design." *IEEE J. of Solid-State Circuits*, Vol.33, No.4, pp.515-534, Apr.1998.

[4.34] Seng-Pan U, R.P.Martins, J.E.Franca, "A 120 MHz SC 4th-Order Elliptic Interpolation Filter with Accurate Gain and Offset Compensation For Direct Digital Frequency Synthesizer," in *Proc. The First IEEE Asia-Pacific Conference on ASICs (AP-ASIC'99)*, pp.1-4, Aug.1999.

[4.35] R.E.Crochiere, L.R.Rabiner, *Multirate Digital Signal Processing*, Prentice-Hall, Inc., NJ, 1983.

[4.36] J.E.Franca, R.P.Martins, "IIR Switched-Capacitor decimator building blocks with optimum implementation," *IEEE Trans. Circuits and Systems*, Vol. CAS-37, No.1, pp.81-90, Jan. 1990.

[4.37] R.P.Martins, J.E.Franca, F.Maloberti, "An optimum CMOS Switched-Capacitor antialiasing decimating filter," *IEEE J. Solid-State Circuits*, Vol.28 No.9, pp.962-970, Sep. 1993.

Chapter 5

DESIGN OF A 108 MHZ MULTISTAGE SC VIDEO INTERPOLATING FILTER

1. INTRODUCTION

A high-performance and economic solution for digital video in modern consumer and professional applications entails a high integration of the large digital system with traditionally external analog interfaces on a single chip, like DVD players, TV-output in DVD-equipped PCs, PC multimedia video editing systems, digital set-top boxes, digital still cameras, video phones, as well as studio and broadcast video systems and others [5.1, 5.2, 5.3, 5.4, 5.5, 5.6, 5.7, 5.8, 5.9]. In such systems, the digital YCrCb (4:2:2) 8-bit or 16-bit component video stream inputs are converted into standard analog composite (NTSC or PAL) or S-video outputs, and a post signal restitution or anti-imaging filter is necessary to smooth the DAC outputs by attenuating the images from the inherent sampling process in digitizing analog video. However, as shown in Figure 5-1(a), many currently available video encoders in the market still require an off-chip passive inductive-capacitive (LC) filter for this post reconstruction filtering [5.1, 5.3, 5.7, 5.8]. The implementation of high-order monolithic CT filters, together with phase-equalization, for wideband video applications is still not straightforward and cost-efficient due to the limitation of the inherently inaccurate time-constant in current IC technology [5.10, 5.11, 5.12, 5.13, 5.14, 5.15, 5.16]. In addition, a $\sin x/x$ correction is usually required prior to the DAC in order to correct for rolloff due to the S/H effects at lower sampling rate, e.g. 2.1 dB roll-off at 13.5 MHz [5.1, 5.3, 5.7, 5.8].

The analog multirate video scheme [5.17, 5.18, 5.19, 5.20] as shown in the Figure 5-1(b) by inserting a multistage 8-fold FIR linear-phase SC interpolating filter, generates CT sampled-and-held signal at 108 MHz output sampling rate. This reduces the speed of the video DSP and DAC or simplify the CT filter to only a very relaxed 1st-order (e.g. -3dB-frequency can be varied within ±20% around the nominal 18 MHz), and at the same time eliminates the sin$x$/x correction (the roll-off at 5.5 MHz with 108 MHz is as small as 0.037dB), as well as the phase equalizer stage, thus leading to a low-cost full single-chip alternative.

This chapter presents an optimum design and realization of this high-frequency 8-fold multistage SC FIR interpolating filter for NTSC/PAL digital video according to CCIR-601 standards [5.1, 5.3, 5.7, 5.8, 5.21, 5.22], i.e., equi-ripple gain (< ±0.25 dB) & linear phase characteristics (< ±10 ns group delay variation) in 5.5 MHz passband with typically a 40 dB stopband attenuation.

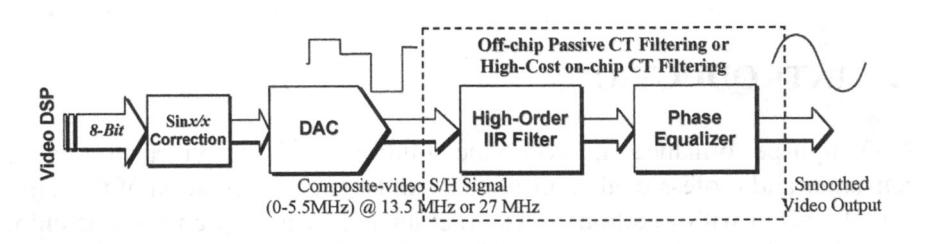

(a)

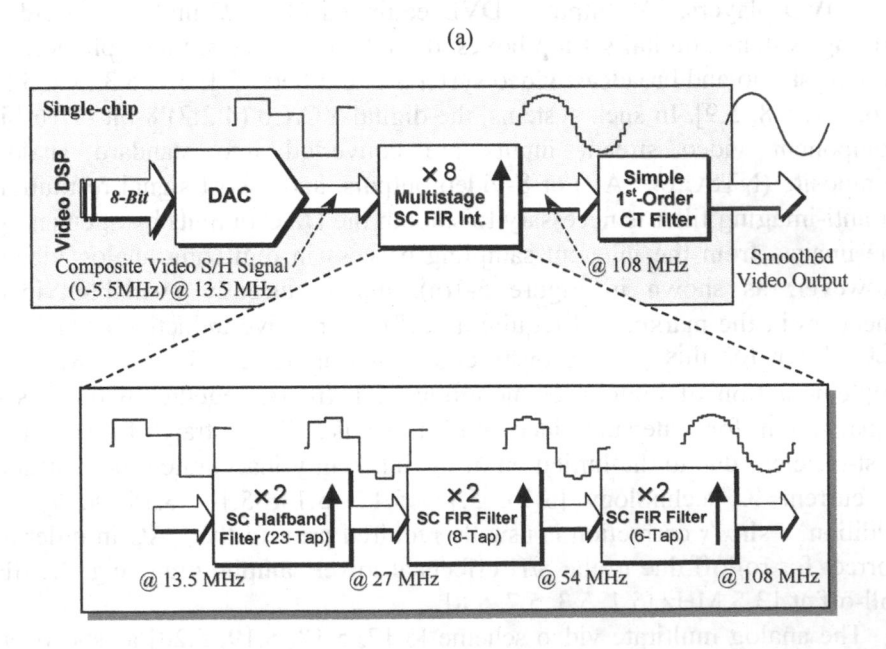

(b)

Figure 5-1. (a) Traditional (b) Multirate alternative for digital video restitution system

2. OPTIMUM ARCHITECTURE DESIGN

2.1 Multistage Polyphase Structure with Half-Band Filtering

In order to use only a very simple and less stringent 1st-order post CT filter, the sampling rate must have 8-fold increase. However, if a single stage design is employed, this large interpolation factor will lead to impractical high FIR filter length and coefficient spread, i.e. 76 and 2200. Multistage implementation, with a series of interpolation stages with relatively small rate change ratio will exhibit greater efficiency in practice when compared with a single stage implementation in terms of filter order and spread, similar to those in multirate digital interpolation [5.23, 5.24, 5.25, 5.26]. As shown in the Figure 5-2, this 8-fold interpolation is implemented in a cascade 3-stage design, and the signal images sampled at 13.5 MHz will be attenuated by these multistage filter stage by stage, and the sampling rate will also be double for each stage. Therefore, in this arrangement, the total filter length is halved to only 23+8+6=37 and maximum spread is 114 (finally, to only less than 20 considering the proposed SC spread reduction technique presented next), and thus directly resulting in considerable savings in silicon and power consumption. More importantly, the dependency of system performance to the capacitance ratio sensitivity will be substantially relaxed with the multistage implementation, because for the original 76-tap design with 0.5 % Gaussian standard deviation of coefficient ratio, the achievable minimum stopband attenuation from (3.6) is about 36 dB which is out of the design specifications, even without counting any other nonideal effects of SC implementations.

In addition, for further optimum realization, half-band filtering techniques [5.23, 5.24, 5.25, 5.26, 5.27], which are specialized for multirate filtering with 2-fold sampling rate alteration ratio, will be efficiently used in the first stage that is the most stringent stage among all others. The main advantage of half-band filter is that almost half of impulse response coefficients are zero which reduces not only the SC branches and capacitance area but also, and relatively, the sensitivity problem which is a critical limitation of long-tap analog transversal filter [5.28, 5.29]. Hence, a 23-tap FIR half-band filtering function is obtained for the first stage with sampling rate increase from 13.5 MHz to 27 MHz, as shown in Figure 5-1.

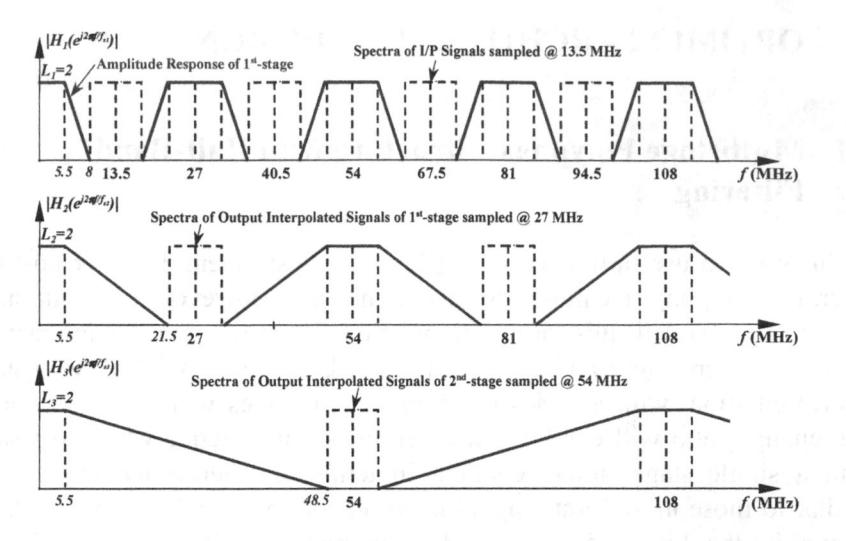

Figure 5-2. 3-stage implementation of 8-fold interpolating filter for digital video restitution system

Although the 2nd- and 3rd-stage, that all have interpolation factor of 2, can use the half-band filtering functions, the system function with even-number order has been employed, i.e. 8- and 6-tap FIR functions for 2nd- and 3rd-stage with 27 MHz to 54 MHz and then 54 MHz to 108 MHz rate increase, respectively. Such solutions will help to boost the high-speed capability with reduced power consumption, as well as improved accuracy for these two stages, as will be analyzed next. Moreover, the half-band ADB polyphase structures [5.30] and direct-form polyphase structures with rotating-switching parallel coefficients [5.31] are employed for the 1st and the last 2 stage filters.

2.2 Spread-Reduction Scheme

One critical problem of analog SC FIR filter is the capacitance spread which is normally very large especially for narrowband filtering. In this design, the spread has been optimized in a system-level from original more than 2000 for a single stage to only maximum of 114, 32 and 26 for 1st-, 2nd and 3rd-stage realization. Especially, the spread in each stage can be further halved to only 57, 16 and 13, respectively, according to the improved analog polyphase structure (divided by interpolation factor L) [5.29]. However, the maximum will be close to 11.4 pF for a unit capacitance of 200 fF that is still impractical for implementation at such high frequency. T-network scheme [5.32] can reduce the spread but at the expense of requiring higher DC amplifier gain and suffering parasitics.

Although some techniques have emerged for SC circuits with very large time constant [5.33, 5.34, 5.35, 5.36, 5.37, 5.38], here we proposed another novel and efficient technique, namely, two-step summing approach, for further spread-reduction. For simplicity, this 2-step summing approach is described based on its SC implementations, as shown in Figure 5-3(a), (b) and (c), respectively, for one higher-speed opamp, double-sampling and autozeroing realizations (Note that the SSC can be applied in both Figure 5-3(a) and (b) for achieving CDS). Its operation is described as follows: the coefficients in 1st-stage are divided into Group A: $h_0(h_{22})$– $h_6(h_{16})$ and Group B: $h_8(h_{14})$– $h_{10}(h_{12})$ which will be normalized separately with their own summing capacitor C_{sumA} and C_{sumB} and their charge transferring will be accomplished in two steps at successive phases A and B respectively, and the summed output in phase A will be transferred to C_{sumB} by an extra C_{uB} in phase B together with the charges from Group B capacitors. Thus, the implemented coefficients for Group B will be simply C_{Bi}/C_{sumB}, while for Group A, it becomes $(C_{Ai}/C_{sumA}) \cdot (C_{uB}/C_{sumB})$. To make an integer ratio of C_{uB} and C_{sumB} for good matching leads to the sensitivity of overall capacitance ratio equivalent to the original one. In this realization, the spread can be reduced to only 8 for 1st-stage, although 10.3 is finally adopted to increase the feedback factor for relaxing OTA speed with dynamic range scaling for two phase outputs. Thus, about 72% reduction in total capacitance area is achieved by the above spread reduction scheme, and more than 30% is further saved by the coefficient-sharing technique that will be discuss later. Note that there is a half period delay between the sampling instant of group A and B SC branches that needs to be arranged during the signal sampling.

2.3 Coefficient-Sharing Techniques

One advantage of FIR linear phase filtering in terms of implementation is the feasibility of the sharing of symmetrical coefficients. Unlike in digital signal processing, it was not achievable previously in pure SC circuits (unless for the semi-analog SC FIR structures [5.39] which is not suitable for high-speed filtering) due to the impossibility of analog summing by only one SC branch. As both positive and negative outputs are available in fully-differential structures, we propose here a simple and elegant solution for sharing coefficients by just subtracting one positive version output signal for another but in negative version that will have same tap weight from distinct delay stages, instead of signals summing. Such subtraction can basically be

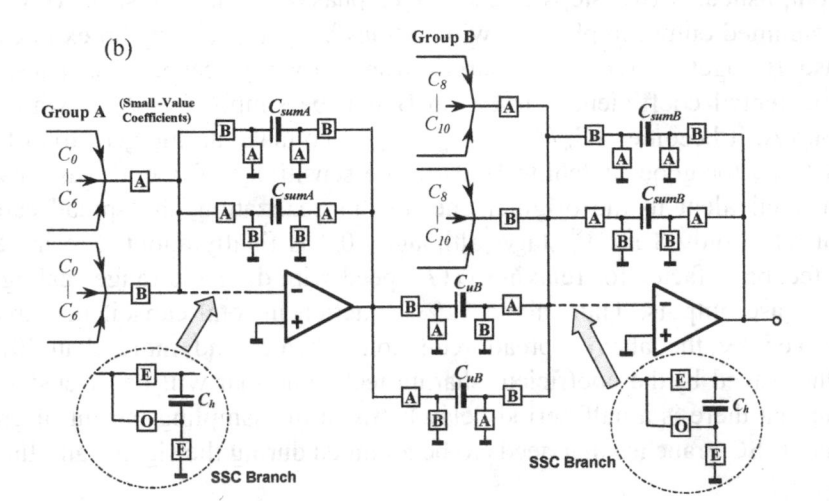

Figure 5-3. (a) One-opamp scheme (b) Double-sampling scheme (c) Autozeroing scheme for spread-reduced two-step summing technique

implemented by SC branches in either an instantaneous-adding or subsequent-adding approaches as shown in Figure 5-4(a) and (b). The latter one is preferred as it is insensitive to parasitics as the former does, and it can also eliminate the signal-dependent charge injection & clock feedthrough errors by using bottom-plate sampling that is not applicable to the former. Note that the latter will subtract two inputs in two consecutive phases whose charge transferring can be described as

$$\Delta Q(z) = C \cdot \left[V_{in1}(z)z^{1/2} + V_{in2}(z) \right] \tag{5.1}$$

where the extra $z^{-1/2}$ delay can be embedded in the delay line during the signal sampling.

(a)

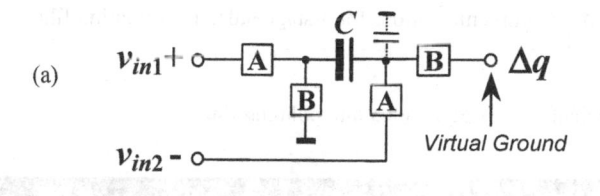

(b)

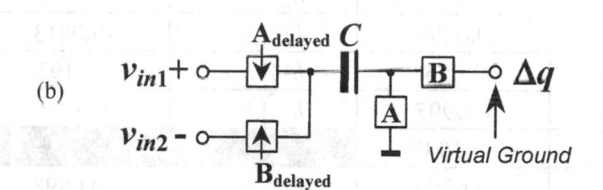

Figure 5-4. (a) Instantaneous-adding (b) Subsequent-adding SC subtraction branches using Coefficient- Sharing Technique

Such novel coefficient-sharing leads to, not only a substantial reduction in total capacitor area which normally dominates the total chip area, but also, more importantly, to a large improvement in capacitance ratio matching and sensitivity. Moreover, it also reduces the total capacitive loading and increases the feedback factor to output accumulator opamp, which in turn, improves the achievable speed of summing circuit and decreases the required power consumption.

3. CIRCUIT DESIGN

The SC circuit implementation for the 3-stage is summarized in the Figure 5-5 [5.40] and the implemented FIR coefficients are listed in Table 5-1. The following sessions will present each stage in detail.

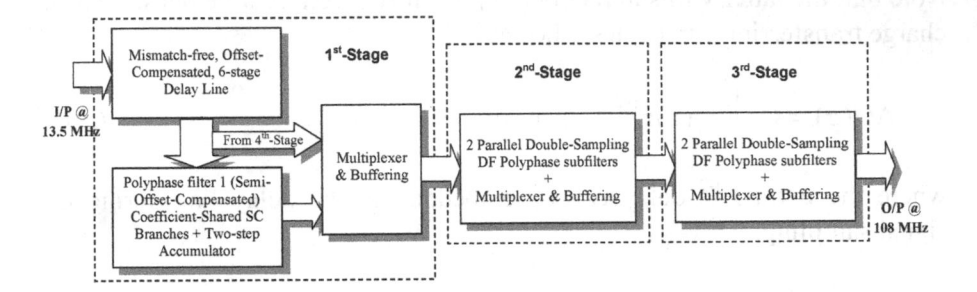

Figure 5-5. SC implementations for 3-stage video interpolating filter

Table 5-1. FIR Coefficients for 3-stage video interpolating filter

1st-Stage		2nd-Stage	
$h_{0,1} \mid h_{22,1}$	-0.0177	$h_{0,2} \mid h_{7,2}$	0.0625
$h_{2,1} \mid h_{20,1}$	0.0299	$h_{1,2} \mid h_{6,2}$	-0.0813
$h_{4,1} \mid h_{18,1}$	-0.0554	$h_{2,2} \mid h_{5,2}$	0.3192
$h_{6,1} \mid h_{16,1}$	0.0997	$h_{3,2} \mid h_{4,2}$	0.9375
$h_{8,1} \mid h_{14,1}$	-0.1947	3rd-Stage	
$h_{10,1} \mid h_{12,1}$	0.6306	$h_{0,3} \mid h_{5,3}$	-0.0898
$h_{11,1}$	1	$h_{1,3} \mid h_{4,3}$	0.1475
others	0	$h_{2,3} \mid h_{3,3}$	0.8462

3.1 1st-Stage

The 1st-stage is the most critical stage requiring a high-order (23-tap) function due to its sharp-selectivity frequency response. ADB delay-line-based FIR structure specialized for half-band filtering [5.26, 5.27], which contains only one polyphase subfilter, is used. However, in order to reduce the accumulated errors along the delay line especially the offset errors which will lead to fixed pattern noise, either autozeroing or CDS techniques is necessary for ensuring the pattern noise below signal in the order of 45 dB. Besides, the minimization of the number of ADB's is also crucial for better noise and especially less power consumption. The final required ADB number in the delay line has also been optimized from the original number close to 20 to only 6, due to the use of the rotating-switch multi-unit delay scheme.

The coefficient-sharing and two-step summing techniques have also been applied in this stage, leading to the final circuit presented in Figure 5-6(a) where the multi-unit AZ SC delays are derived from Figure 4-3(c) and Figure 4-5(a) in previous chapter and thus not shown here for simplicity. For clarity, the following summarizes the employed techniques in this stage.

- Half-band ADB polyphase structure
- Mismatch-Free and Multi-unit ADB
- Semi-Autozeroing (or EC/P-CDS)
- Coefficient-sharing technique
- Two-step summing technique

After an optimization among the number of delay blocks and coefficient sensitivity taking into account the arrangement of coefficient-sharing, spread-reduction, as well as AZ scheme, only h_4 & h_{18}, h_6 & h_{16}, h_8 & h_{14}, and h_{10} & h_{12} have been efficiently shared except h_0 & h_{22}, h_2 & h_{20} in which some are indeed implemented by parallel/rotating switching SC branches for achieving longer delay, as shown in Figure 5-6(a). This is due to the fact that they have the smallest value and, particularly, less sensitivity among all others, thus no big impact on silicon area and sensitivity. Especially, by taking advantage of this low-sensitivity property, and for saving more delay block or power and for achieving an even-number of delay, h_2, h_{20} and h_{22} are implemented by TSI branches that are not able to store the offset error during the signal sampling, and thus rendering offset error. However, the resulting pattern-noise tone due to this so-called semi-autozeroing circuit are still all >50 dBc from simulations, which is still within the specifications. Furthermore, the sole mid coefficient h_{11} in polyphase filter 1, which is the most sensitive one that affects especially stopband in half-band filter, has been adjusted to unity and was obtained directly from the mismatch-free delay line for eliminating the capacitance ratio deviation. The two-step summing scheme is used here for the two group of coefficients, i.e. Group A: $h_0(h_{22})$– $h_6(h_{16})$ and Group B: $h_8(h_{14})$– $h_{10}(h_{12})$.

For comparison purposes, another version of the 1st-stage implemented by EC/P-CDS techniques is also presented in Figure 5-6(b) where the ADB cells are obtained from the Figure 4-3(g) and (j) and Figure 4-5(d) and the one-time-shared EC/P-CDS accumulator is modified from Figure 4-6(e) using SSC GOC techniques [5.30]. More importantly, the power analysis for these two circuits is illustrated in the Table 5-2. The results validate that the special low gain requirement of opamps for EC/P-CDS techniques trades off with the consumed power dissipation being in this case much higher and almost 3 times than that for AZ.

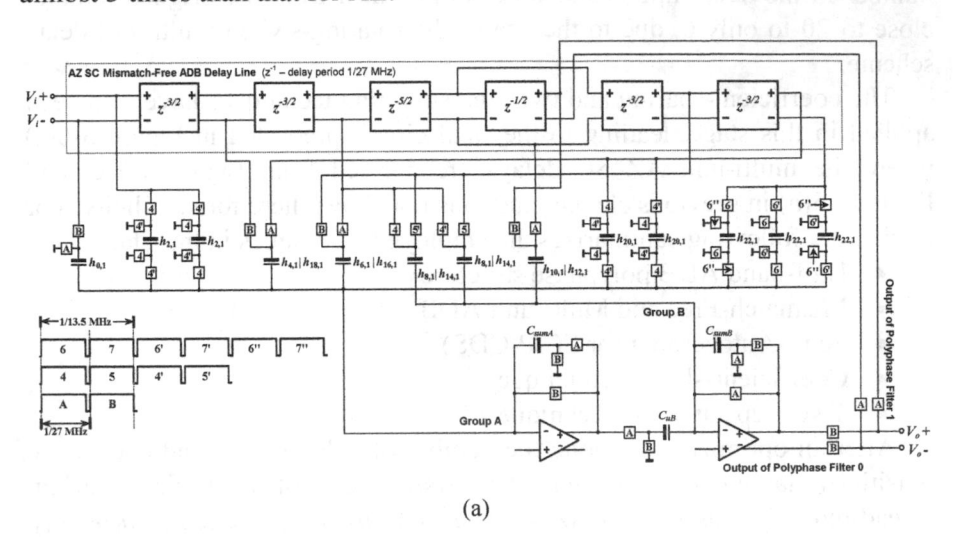

(a)

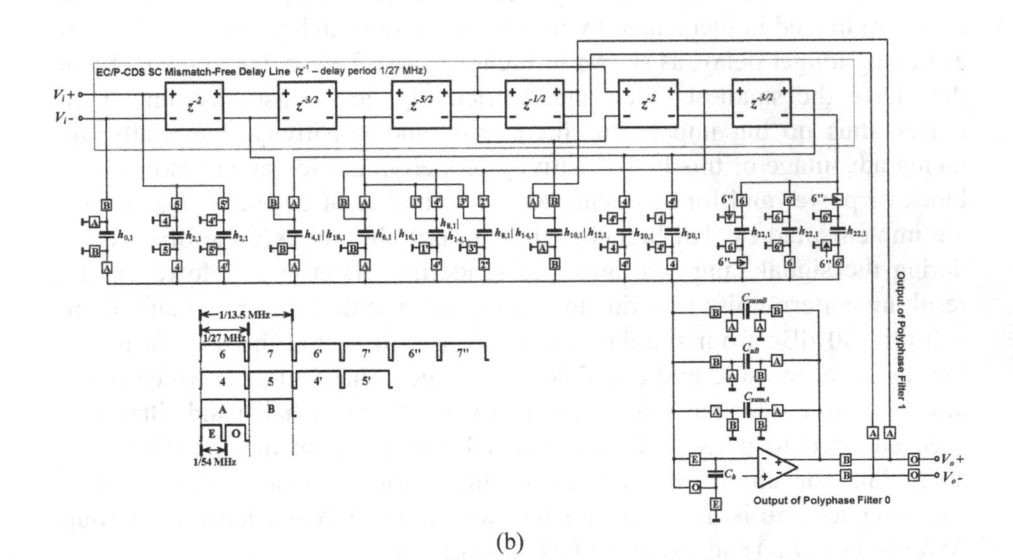

(b)

Figure 5-6. (a) AZ (b) EC/P-CDS SC implementations for the 1st-stage

Table 5-2. Power comparisons for 1st-stage in AZ of Figure 5-6(a) and EC/P-CDS of Figure 5-6(b)

	AZ - ADB		AZ - ACCU		EC/P-CDS - ADB		EC/P-CDS - ACCU	
Phase	Cali.	O/P	Cali.	O/P	Cali.	O/P	Cali.	O/P
Settling Time	35 ns	35 ns	35 ns	35 ns	16.5 ns	16.5 ns	16.5 ns	16.5 ns
Max. $V_{o\text{-}step}$	0.7 V	0.7 V	0.7 V	0.7 V	1.3 V	0.7 V	1.3 V	0.7 V
SR	120 V/µs	120 V/µs	120 V/µs	120 V/µs	390 V/µs	210 V/µs	472 V/µs	255 V/µs
FB Factor	1	0.32	1	0.46	0.31	0.25	0.25	0.16
Equivalent C_{Ltot}	2.6 pF	4 pF	9.3 pF	4.8 pF	1.35 pF	5 pF	6 pF	5.9 pF
g_m	0.6 mS	3 mS	2.3 mS	2.6 mS	1.64 mS	7.6 mS	10 mS	16 mS
I_{SS}	0.31 mA	0.48 mA	1.12 mA	0.58 mA	0.53 mA	1.52 mA	2.82 mA	3.73 mA
Total Power	14 mW (8 opamps)				39 mW (7 opamps)			

Note: 2 ns is allocated for Non-overlapping phase gap and delay time for bottom-plate sampling technique
The highest g_m and I_{SS} are presented for the opamps in ADB's and accumulators

3.2 2nd- and 3rd-Stage

The relatively lower filter order feature for 2nd- and 3rd-stage allows the possibility of elimination of the ADB delay line structures, which increase the sensitivity to the opamp offset and gain errors. The direct-form polyphase structures with multi-unit delay by parallel rotating-switching and double-sampling accumulation are feasible here without extra offset compensation phases, so the opamps can have full wide input sampling period for settling so as to minimize the power consumption and enlarge the design headroom of the high frequency opamp which is the more important for these two stages. Especially, the even-number filter order is employed, e.g. 8 and 6 for 2nd- and 3rd-stage respectively, so that each of 2 polyphase filters ($L=2$) will contain same tap number and weight, thus contributing with a more balanced error factor to two paths for reducing, relatively, the path offset mismatches. In this way, the overall offset errors are more dependent on the random opamp DC offset mismatch that can be improved by using careful layout techniques.

The simplified circuit schematic for these two stages is presented in Figure 5-7. As presented, $h_{4,2} - h_{7,2}$ as well as $h_{4,3} - h_{5,2}$ are implemented by rotating-switching SC branches for achieving a longer delay. Individual output accumulator with double-sampling approach and the mismatch-free SC multiplexer, discussed in Chapter 2 / Session 4.2, are employed here. It is worth to point out that the multiplexers between stage 1 and 2 (not shown in the figure for simplicity), as well as stage 2 and 3 mainly function as an buffer interface for providing enough driving capability, although they can be replaced by the next stage SC branches with proper sampling arrangement to the accumulator output. This buffering is necessary for significantly lessening the capacitive burden of the accumulators that are the vital part of the circuit and extensively relate to the system response accuracy. From the calculations the power budget in this way is even lower than that without multiplexers. Especially, this type multiplexers introduce less gain and offset mismatches and suitable for high-frequency operation aforementioned in Chapter 2 / Session 4.2.

The power assignment for these two stages is presented in Table 5-3. The highest power-consuming element is the opamp 3 in the output multiplexer of the 2nd-stage, due solely to the need to drive large capacitive loading imposed by all 3rd-stage coefficient capacitors. The total expected power is about 30 mW for two stages.

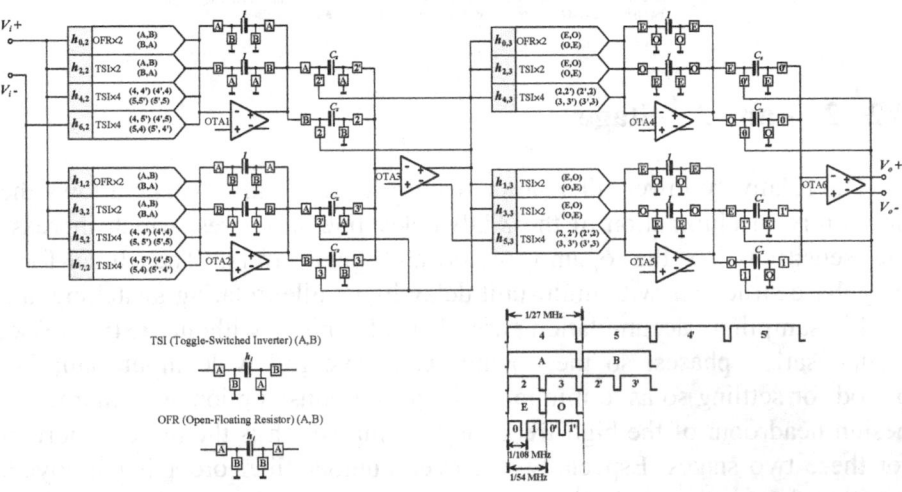

Figure 5-7. Simplified SC implementations for the 2nd- and 3rd-stage

Table 5-3. Power analysis for 2nd- and 3rd-stage

opamp	2nd-Stage			3rd-Stage	
	Mux	1 & 2	3	4 & 5	6
Settling Time	35 ns	35 ns	16.5 ns	16.5 ns	7.25 ns
Max. $V_{o\text{-}step}$	1.2 V	1.2 V	0.7 V	0.7 V	0.4 V
SR	220 V/μs	220 V/μs	300 V/μs	250 V/μs	320 V/μs
FB Factor	0.53	0.34	0.54	0.36	0.58
Equivalent C_{Ltot}	9 pF	4.3 pF	7.5 pF	4 pF	3.2 pF
g_m	4 mS	3 mS	6 mS	5.4 mS	6 mS
I_{SS}	2 mA	1 mA	2.2 mA	1.1 mA	1.2 mA
Total Power		19 mW		11 mW	

Note: 2 ns is allocated for Non-overlapping phase gap & delay time for bottom-plate sampling technique

The total estimated noise performance for this 3-stage filter have also been calculated with transistor-extracted parameters according to the analysis methodology aforementioned, and the worst-case total noise is less than 300 μV_{rms}, so considering a 1 V_{p-p} input signal, it leads to the SNR greater than 60 dB which is satisfactory for our specifications.

3.3 Digital Clock Phase Generation

Digital phase generation network whose block diagram is shown in Figure 5-8 needs to provide in total 44 phases including fall-time delayed version of the main clock for the bottom-plate sampling. Robust synchronization among submaster clocks and all multiple phase outputs with different periods was achieved by using optimized logic structures, so that the systematic mismatches in rising edges could be only happen in the phase-width control and buffering circuitry. For simplicity, only the submaster clock generation and phase-width control circuitry are shown in Figure 5-9(a) and (b), respectively. The circuit for Figure 5-9(a) provides 3 submaster clock synchronized at 108 MHz, 54 MHz and 27 MHz with a signal clock input, while the phase-width control cell, which is used for all phases from logic encoder, generates two rising-edge synchronized phases but with different falling-edge controls, as shown in Figure 5-9(b).

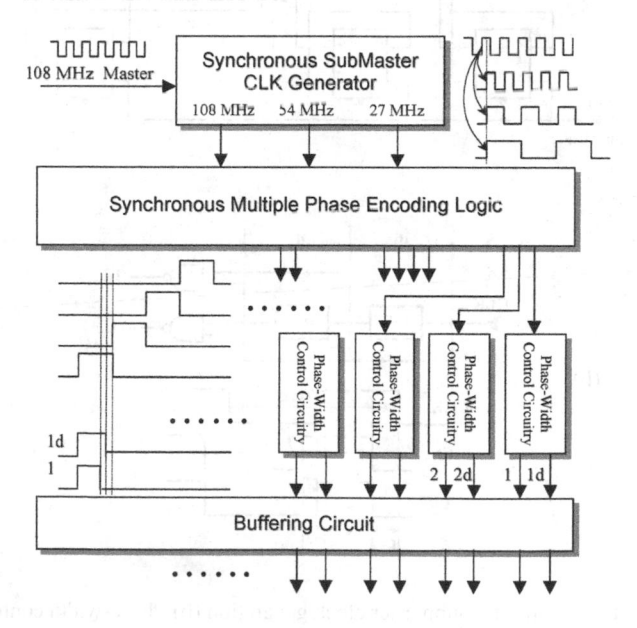

Figure 5-8. Multiple phase generation block diagram for multistage SC video interpolating filter

In order to evaluate the influence of the timing-skew errors in the clock phase, the Figure 5-10 presents the 100-time Monte-Carlo simulated mean value of SNR and SFDR of a 5.5 MHz signal sampled at 108 MHz (uniformly sampled value but with non-uniformly S/H output) versus different path number and standard deviation of the timing skew errors. For ensuring SNR>45 dB, the timing skew error must be smaller than 2-3 hundred pico second. From the overall clock phase generation structure, the main contribution of the random mismatches happens in the phase-width control, as well as in the buffering circuitry where the mean value of phase skew deviation for these blocks, by the Monte-carlo HSPICE simulations is smaller than 100 ps.

The simulated total dynamic power for the digital phase generation is less than 30 mW where a large portion of that is dissipated by the buffering circuitry for driving all the switches.

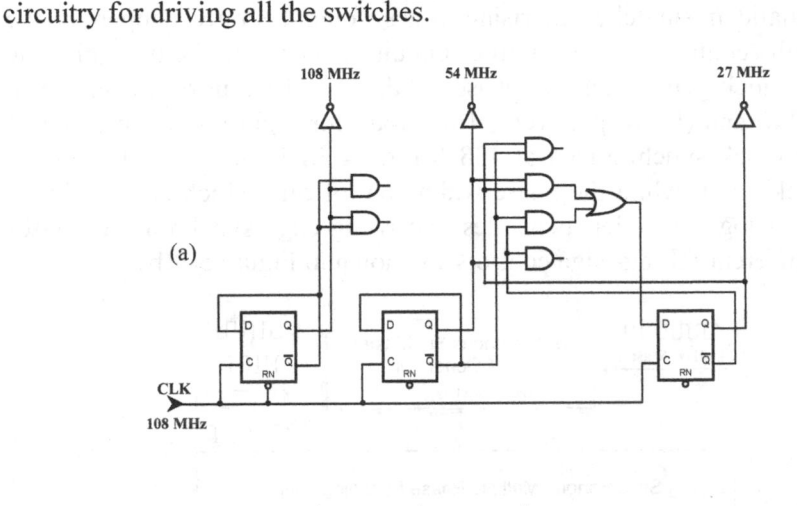

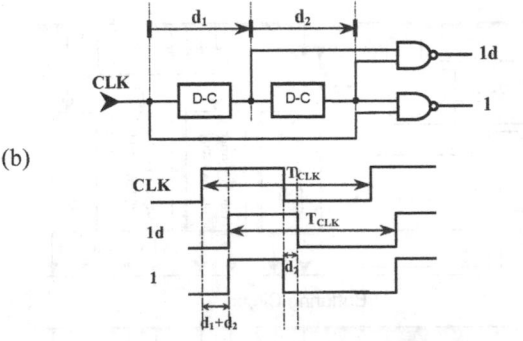

Figure 5-9. (a) Synchronize Submaster clock generation (b) Phase-width controls circuitry

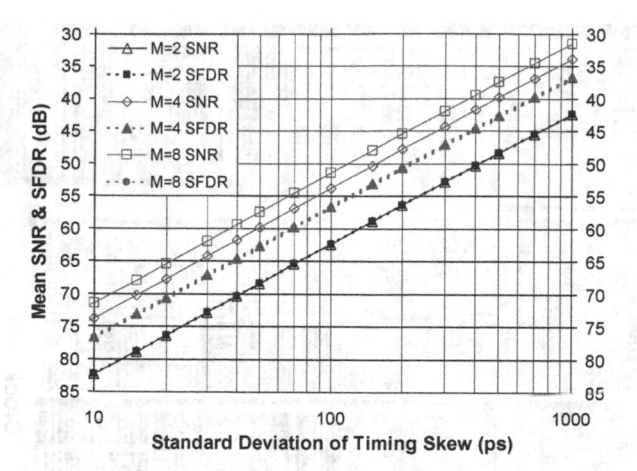

Figure 5-10. SNR and SFDR Mean vs. timing-skew errors (100-time Monte-Carlo) (f_{in}=5.5 MHz, f_s=108 MHz)

4. CIRCUIT LAYOUT

The circuit was designed by employing a 0.35 μm double-poly triple-metal CMOS technology. For minimization of the power at higher operation speed, the single-stage telescopic OTA structure, with Miller-effect cancellation transistors and wide-swing, internal-biasing for the cascode transistors was adopted here. The common-mode voltage is stabilized at 1.1 V by a dynamic SC common-mode feedback circuit. The fastest opamp achieves 70 dB gain, 302 MHz GBW with 64° PM, and 6.5 ns settling time and consumes 6.6 mW of power from layout-extracted Cadence simulation results.

Common-centroid plus dummy periphery and mirror-symmetrical arrangements have been employed for the mismatch-sensitive analog parts, like opamp differential pair, and critical capacitor group. One analog and one digital VDD supply are used with a shared ground. On-chip MOS capacitors between VDD and ground are inserted in any empty area for having enough decoupling capacitance. The overall circuit layout is shown in Figure 5-11 with an active area about 3.3 mm^2 including both analog and digital parts.

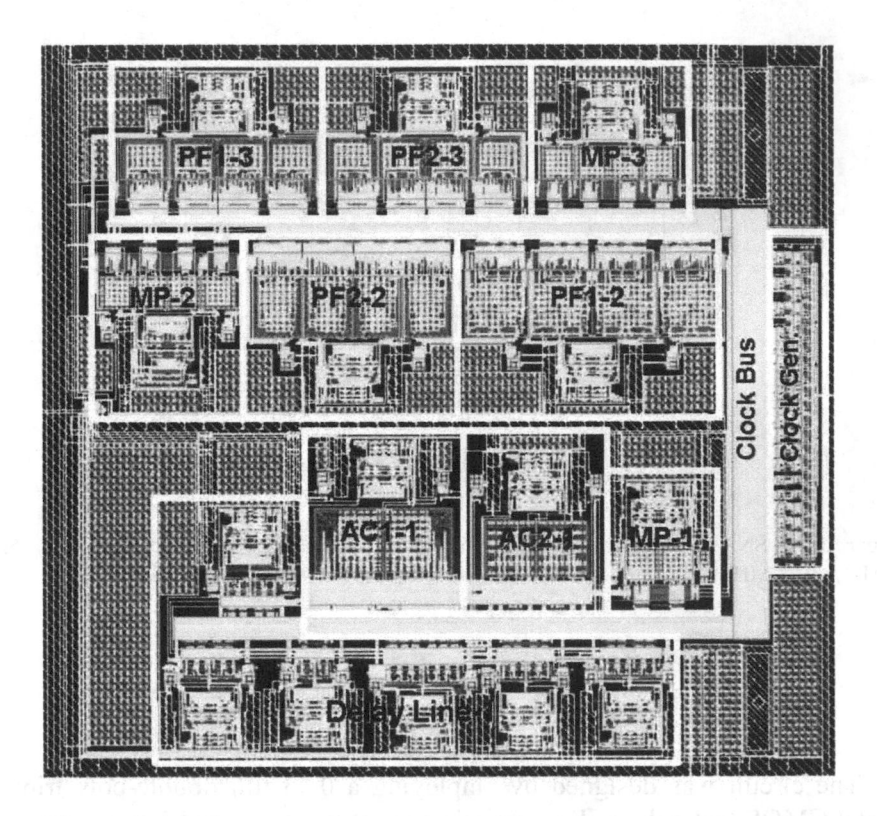

Figure 5-11. Circuit layout for 3-stage 8-fold SC interpolating filter (AC-Accumulator, PF-Polyphase Filter, MP-Multiplexer)

5.   SIMULATION RESULTS

5.1 Behavioral Simulations

To evaluate the random capacitance coefficient mismatch errors, Monte-Carlo amplitude simulation for capacitance ratio being independent zero-mean Gaussian random variables with the deviation within 1.5 % has been performed and the results are shown in Figure 5-12. It shows that the upper and lower bound deviation in passband are within the desired ±0.25 dB including the output S/H shaping effect, and the unwanted image bands located at 13.5 MHz, 27 MHz, 54 MHz have been all attenuated by the 1st-, 2nd- & 3rd-stage with a value subsequently greater than 40 dB.

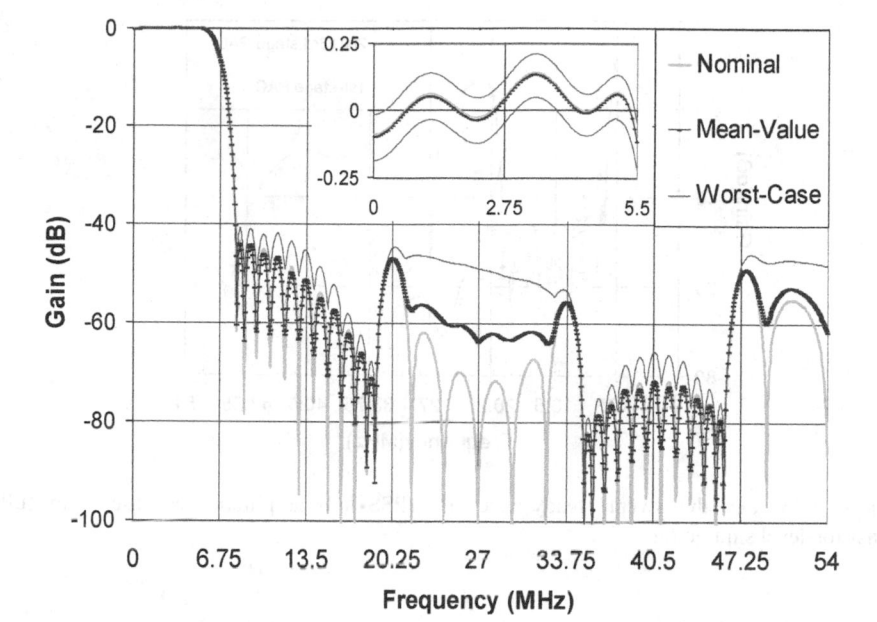

Figure *5-12*. Monte-Carlo amplitude response simulation (500-time, $\sigma = 0.5$ %)

5.2 Circuit-Level Simulations

The circuit was also fully verified by the transistor-level and parasitics-involved layout-extract post simulations. Figure 5-13 shows that the amplitude response for the 1st and last 2 stages (2nd+3rd) from the transistor-level Periodic Swept Steady-State AC (PSS-AC) analysis using Spectre simulator in Cadence package, which shows that the circuit meets well the specifications. Figure 5-14 presents the spectrum of 108 Msample/s output signal from this 3-stage filter with a 5 MHz, 1.3Vp-p, 13.5 Msample/s input. The result is obtained from FFT of one worst-case transistor-level transient simulation with opamp transistors sized by Monte-Carlo mismatches at a maximum of 6 %. It is obvious that the unwanted images of input baseband signal and the fixed pattern noise tones around 13.5, 27 and 54 MHz have been supressed below 45 dB, and the rest spurs are the frequency translated-images of the harmonics sampling at different input & output rates, and their multiples, which result in a Total Harmonic Distortion (THD) still below – 60 dB.

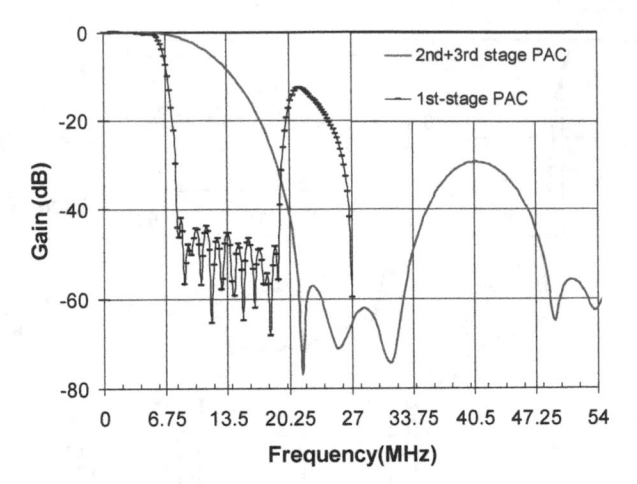

Figure 5-13. Periodic swept steady-state AC (PSS-AC) amplitude response from full transistor-level simulation

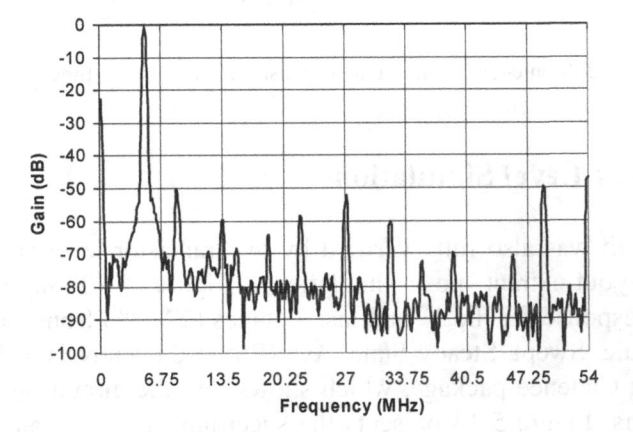

Figure 5-14. Spectrum of 5 MHz @ 108 MHz output signal from the worst-case transistor-level simulation

In addition to the verification by the LVS (Layout versus Schematic) CAD tool to the complete layout, the FIR impulse responses to each stage and the overall system have been performed for validating the practical circuit effectiveness, by the parasitic-involved post-layout transient simulation, but with a reduced computation time. Figure 5-15(a), (b) and (c) are the achieved symmetrical FIR impulse transient responses of 1st 23-tap half-band, 2nd 8-tap + 3rd 6-tap stage and the whole 3-stage system, respectively.

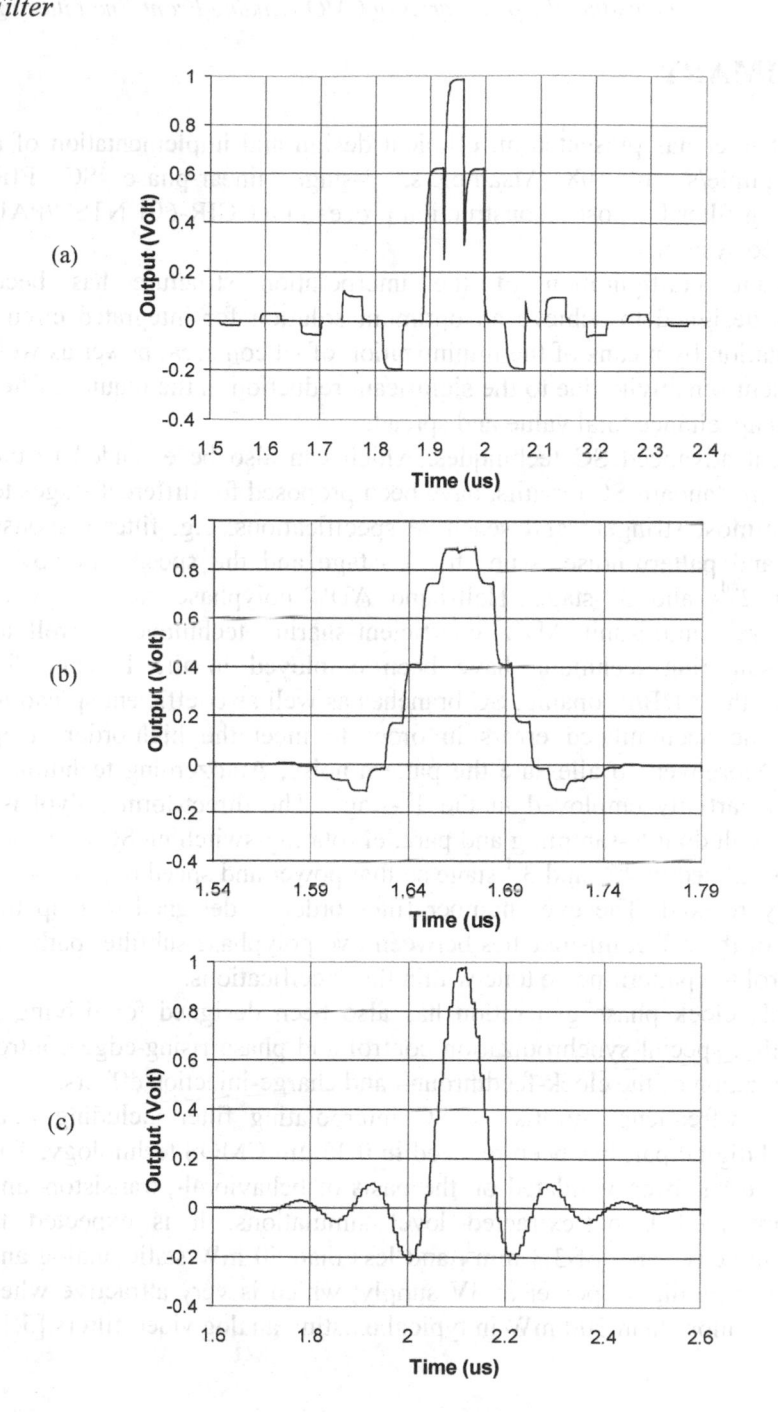

Figure 5-15. Impulse transient response from parasitic-involved layout-extracted simulation
(a) 1st-stage (b) 2nd+3rd stage (c) overall 3-stage

6. SUMMARY

This chapter has presented an efficient design and implementation of a 13.5 Msample/s to 108 Msample/s, 3-stage linear-phase SC FIR interpolating filter for post reconstruction proces in a CCIR-601 NTSC/PAL digital video system.

Multistage configuration of the interpolation structure has been efficiently designed to achieve an optimum solution for integrated circuit implementation by means of the minimization of silicon area, power as well as coefficient sensitivity due to the significant reduction in the required filter order and capacitance total value and spread.

Different advanced SC techniques, which can also be extended to the utilization in standard SC circuits, have been proposed for different stages to relax their most stringent and sensitive specifications, e.g. filter response accuracy and pattern-noise issues for 1st-stage and the speed and power issues for 2nd- and 3rd-stage. Half-band ADB polyphase structure with mismatch-free, multi-unit ADB, coefficient-sharing technique as well as two-step summing technique have been employed in the 1st-stage for minimizing the ADB/or opamp, SC branches as well as coefficient spread to minimize the accumulated errors in order to meet the high-order steep response. Moreover, to alleviate the pattern-noise, Autozeroing techniques have been partially employed in the 1st-stage. The direct-form polyphase structures with double-sampling and parallel rotating-switching SC branches have been utilized in 2nd- and 3rd-stage so that power and speed requirements are greatly relaxed. The even-number filter order is designed to help the reduction of the offset mismatches between two polyphase subfilter paths so as to control the pattern-noise tone within the specifications.

Multiple clock phase generation has also been designed for driving 3 stages with a special synchronization control and phase rising-edge control for minimization of the clock-feedthrough and charge-injection effects.

This high-frequency multistage SC interpolating filter including both analog and digital part has been realized in 0.35 µm CMOS technology. The performance has been validated on the basis of behavioral-, transistor- and parasitic-involved layout-extracted level simulations. It is expected to consume an active area of 3.3 mm^2, and less than 50 mW static analog and 30 mW average digital power at 3V supply, which is very attractive when compared to more than 300 mW in typical existing analog video filters [5.15, 5.16].

REFERENCES

[5.1] Philips Semiconductors, "SAA7199B, Digital Video Encoder (DENC) Data Sheet," 1996.

[5.2] J.Adélaide, et al, "Communication in a single-chip MPEG2 A/V/G decoder for digital set-top box application," *Proc. European Solid-State Circuits Conference (ESSCIRC)*, pp.348-351, Sep.1996.

[5.3] T.Cummins, B.Murray, C.Prendergast, "A PAL/NTSC digital video encoder on 0.6 μm CMOS with 66 dB typical SNR, 0.4% differential gain, and 0.2° differential phase," *IEEE J. of Solid-State Circuits*, Vol.32, No.7, pp.1091-1100, Jul.1997.

[5.4] S.G.Smith et al, "A single-chip CMOS 306×244-pixel NTSC video camera and a descendant coprocessor device," *IEEE J. of Solid-State Circuits*, Vol.33, pp.2104-11, Dec.1998.

[5.5] H.Sanueli, "Broadband communications ICs: Enabling high-bandwidth connectivity in the home and office," *ISSCC Digest of Technical Papers*, pp.26-28, Feb.1999.

[5.6] M.Harrand, et al, "A single-chip CIF 30Hz H261, H263, and H263+video encoder/decoder with embedded display controller," *ISSCC Digest of Technical Papers*, pp.268-269, Feb.1999.

[5.7] Analog Devices Inc. "ADV7177/8 Integrated Digital CCIR-601 to PAL/NTSC Video Encoder Data Sheet," 1998.

[5.8] Conexant Systems, Inc "BT860/861 Multiport YcrCb to NTSC/PAL Digital Encoder Data Sheet," 1999.

[5.9] D'Luna, et al, "A universal cable set-top box system on a chip," *ISSCC Digest of Technical Papers*, pp. 328-329, Feb.2001.

[5.10] V.Gopinathan, Y.Tsividis, K.Tan, "A 5V 7th-order Elliptic analog filter for digital video applications," *ISSCC Digest of Technical Papers*, pp.208-209, Feb.1990.

[5.11] S.D.Willingham, K.W.Martin, "A BiCMOS low-distortion 8 MHz lowpass filter," *ISSCC Digest of Technical Papers*, pp.114-115, Feb.1993.

[5.12] B.Stefanelli, A.Kaiser, "A 2-μm CMOS fifth-order low-pass continuous-time filter for video-frequency applications," *IEEE J. Solid-State Circuits*, Vol.28, No.7, pp.713-718, Jul.1993.

[5.13] I.Bezzam, C.Vinn, R.Rao, "A fully-integrated continuous-time programmable CCIR 601 video filter," *ISSCC Digest of Technical Papers*, pp.296-297, Feb.1995.

[5.14] S.Barbu, Ph.Gandy, B.Guyot, H.Marie, "8-bit acquisition interface with fully integrated analog filter bank for PAL/SECAM/NTSC standards," *Proc. European Solid-State Circuits Conference (ESSCIRC)*, pp.91-21, Sep.1995.

[5.15] Sang-Soo Lee, C.A.Laber, "A BiCMOS continuous-time filter for video signal processing applications," *IEEE J. Solid-State Circuits*, Vol.33, No.9, pp.1373-1381, Sep.1998.

[5.16] Microelectronics Modules Corporation, "Active Video Filter Product List," 2001.

[5.17] P.Senn, M.S.Tawfik, "Concepts for the restitution of video signals using MOS analog circuits," in *Proc. IEEE International Symposium on Circuits and Systems (ISCAS)*, pp.1935-38, 1988.

[5.18] R.P.Martins, J.E.Franca, "A 2.4μm CMOS Switched-Capacitor video decimator with sampling rate reduction from 40.5 MHz to 13.5 MHz," in *Proc. IEEE Custom Integrated Circuits Conference (CICC)*, pp. 25.4/1 -25.4/4, May 1989.

[5.19] J.E.Franca, R.P.Martins, "Novel solutions for anti-aliasing and anti-imaging filtering in CMOS video interface systems," in *Proc. IEEE Workshop on Visual Signal Processing and Communications*, Taiwan, Jun. 6-7, 1991.

[5.20] J.E.Franca, A.Petraglia, S.K.Mitra, "Multirate analog-digital systems for signal processing and conversion," *Proceedings of The IEEE*, Vol.85, No.2, pp.242-262, Feb.1997.

[5.21] "Encoding parameters of digital television for studios," *CCIR International Radio Consultative Committee Recommendation*, pp.601-602.

[5.22] "Analog filter design for video A/D and D/A converters," *Application Note AN35*, Brooktree Corporation, 1993.

[5.23] R.E.Crochiere, L.R.Rabiner, *Multirate Digital Signal Processing*, Prentice-Hall, Inc., NJ, 1983.

[5.24] Markku Renfors, Tapio Saramaki, "Recursive Nth-band digital filters - Part II: Design of multistage decimators and interpolators," *IEEE Trans. on Circuits and Systems*, Vol.CAS-34, No.1, pp.40-51, Jan. 1987.

[5.25] R.Ansari, Bede Liu - "Multirate Signal Processing," Chapter 14 in *Handbook for Digital Signal Processing*, edited by Sanjit K. Mitra and James F. Kaiser, John Wiley & Sons, Inc., 1993.

[5.26] Seng-Pan U, *Impulse Sampled Switched-Capacitor Sampling Rate Converters*, Master Thesis, University of Macau, 1997.

[5.27] Seng-Pan U, R.P.Martins, J.E.Franca, "A novel Half-Band SC architecture for effective analog impulse sampled interpolation," in *Proc. of IEEE International Conference on Circuits, Electronics, and Systems*, pp.389-403, Sep.1998.

[5.28] A.Petraglia, S.K.Mitra, "Effects of coefficient inaccuracy in switched-capacitor transversal filters," *IEEE Trans. Circuits and Systems*, Vol.38, No.9, pp.977-983, Sep. 1991.

[5.29] Seng-Pan U, R.P.Martins, J.E.Franca, "Improved Switched-Capacitor interpolators with reduced sample-and-hold effects," *IEEE Trans. Circuits and Systems – II: Analog and Digital Signal Processing*, Vol.47, No.8, pp.665-684, Aug. 2000.

[5.30] Seng-Pan U, R.P.Martins, J.E.Franca, "A linear-phase Halfband SC video interpolation filter with coefficient-sharing and gain- & offset-compensation," in *Proc. The IEEE International Symposium on Circuits and Systems (ISCAS)*, Vol.III, pp.177-180, May 28-31, 2000.

[5.31] Seng-Pan U, R.P.Martins, J.E.Franca, "High-frequency low-power multirate SC realizations for NTSC/PAL digital video filtering," in *Proc. The IEEE International Symposium on Circuits and Systems (ISCAS)*, Vol.I, pp.204-207, Sydney, May, 2001.

[5.32] W.M.C.Sansen, P.M.V.Peteghem, "An area-efficient approach to the design of very-large time constants in Switched-Capacitor integrators," *IEEE J. of Solid-State Circuits*, Vol.SC-19, No.5, pp.772-780, Oct.1984.

[5.33] G.E.F-Verdad, F.Montecchi, "SC circuit for very large and accurate time constant integrators with low capacitance ratios," *IEE Electronics Letters*, Vol.22, pp.1025-1027, Aug.1985.

[5.34] Qiuting, Huang, "A novel technique for the reduction of capacitance spread in high-Q SC circuits," *IEEE Trans. Circuits and Systems*, Vol.36, No.1, pp.121-126, Jan. 1989.

[5.35] K.Nagaraj, "Parasitic-insensitive area-efficient approach to realizing very large time constants in Switched-capacitor," *IEEE Trans. Circuits and Systems*, Vol.36, No.9, pp.1210-1216, Sep. 1989.

[5.36] Wing-Hung Ki, Gabor, Temes, "Area-efficient gain-and offset-compensated very-large-constant SC biquads," in *Proc. The IEEE International Symposium on Circuits and Systems (ISCAS)*, pp.29-1029, May, 1993.
[5.37] J.Lin, T.Edwards, S.Shamma, "Offset-compensated area-efficient Switched-Capacitor sum-gain amplifier," in *Proc. The IEEE International Symposium on Circuits and Systems (ISCAS)*, pp.1026-1029, May, 1993.
[5.38] N.A.Radev, K.P.Ivanov, "Area-efficient gain- and offset-compensated very-large-time-constant SC integrator," *IEE Electronics Letters*, Vol.36, pp.394-396, Mar.2000.
[5.39] Qiuting Huang, "Mixed analog/digital, FIR/IIR realization of a linear-phase lowpass filter," *IEEE J. of Solid-State Circuits*, Vol.31, No.9, pp.1347-1350, Sep.1996.
[5.40] Seng-Pan U, Ho-Ming Cheong, Iu-Leong Chan, Keng-Meng Chan, U-Chun Chan, Mantou Liu, R.P.Martins, J.E.Franca, "An SC CCIR-601 video restitution filter with 13.5 Msample/S input and 108 Msample/S output," in *Proc. The 4th International Conference on ASIC (ASICON'2001)*, pp.374-377, Oct. 2001.

Chapter 6

DESIGN OF A 320 MHZ FREQUENCY-TRANSLATED SC BANDPASS INTERPOLATING FILTER

1. INTRODUCTION

Frequency-translated sampled-data filtering was initially introduced for very narrow band filtering [6.1]. Its usage has since then been extended to many more applications, namely: the subsampling architecture with sampled-data filtering, which combines both functions of channel selection and frequency downconversion for radio wireless receiver, is one of the notable examples nowadays [6.2, 6.3, 6.4, 6.5]. This chapter presents a complementary approach combining frequency-band selection with frequency upconversion for using in an 8-bit Direct Digital Frequency Synthesis (DDFS) system. DDFS systems have been increasingly employed in modern wireless communication systems due to their fast frequency switching, high purity, reduced phase noise, and fine frequency steps when compared to conventional PLL-based synthesis techniques [6.6, 6.7, 6.8, 6.9, 6.10, 6.11]. However, one of the most important drawbacks is its power-hungry property due to the digital circuit nature, especially at high frequency.

In the traditional design, as shown in Figure 6-1(a), the ROM-based DDFS digital core containing a phase accumulator together with a phase-to-sine amplitude ROM Look-Up Table (RUT), and the following linear DAC, are targeted to operate at normally 3-4 fold higher frequency than the synthesis signal band for simplifying the post CT smoothing filter, i.e. the DAC operates at 240 Ms/s for desired 56-58 MHz signal band, combined with a 4th-order CT post filter that still needs a complex tuning circuitry for on-chip implementation.

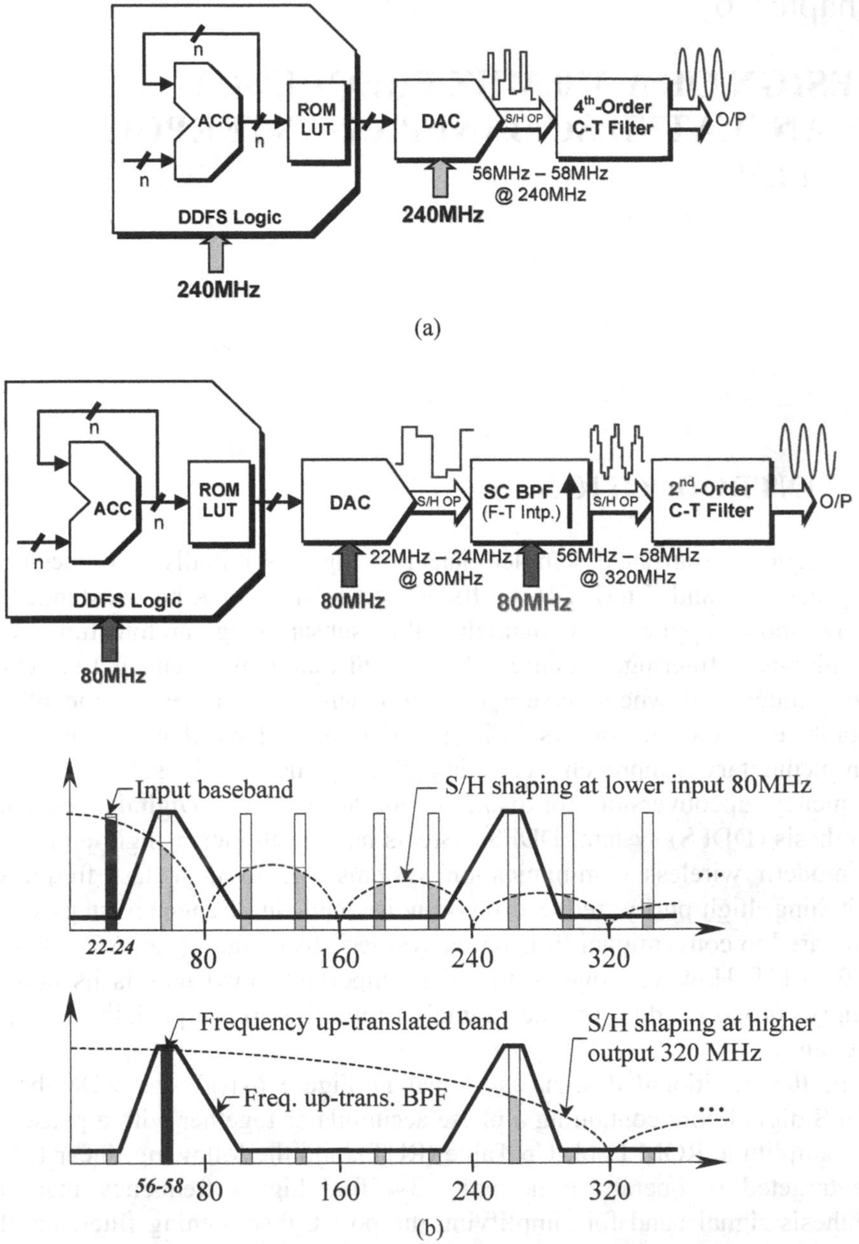

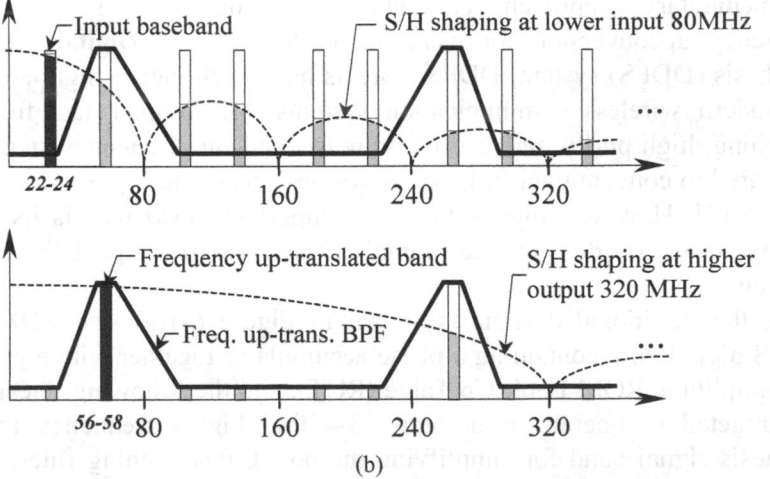

Figure 6-1. (a) Traditional ROM-based DDFS system (b) Proposed DDFS system with frequency-translated SC bandpass interpolation filtering and its signal spectrum

In the proposed approach here, as shown in Figure 6-1(b), an SC bandpass interpolating filter is inserted between the DDFS digital core + DAC and the output CT smoothing filter to translate 22-24 MHz frequency band to 56-58 MHz with an embedded sampling rate increase from 80 MHz to 320 MHz. This allows a 3-fold relaxation in speed to 80 MHz of the DDFS digital core and the DAC with reduced synthesis signal frequency range at 22-24 MHz and also a 2-fold reduction in the post smoothing filter order that is reduced to a simple biquad. From the reported DDFS system operating in the hundred megahertz range, it normally dissipates from hundreds of milli-watt to several watt of power, depending on the system architecture [6.6, 6.7, 6.8, 6.9, 6.10, 6.11]. Thus, a 3-fold reduction in operating speed directly results in a 3-fold power reduction besides other benefits in terms of design complexity, system area and performance. Indeed this also gives a more than 3-fold enhancement in the design headroom for an 80 MHz DAC in contrast with the design of a 240 MHz DAC. Moreover, the power per pole for state-of-the-art CT filter, with cut-off frequency beyond 50 MHz and at least more than 45 dB dynamic range and 1 V_{p-p} output, costs about several tens of milli-watt depending on the filter topology [6.12, 6.13, 6.14, 6.15, 6.16, 6.17, 6.18], so the required power will also be halved in this approach. Moreover, designing a more relaxed CT biquad is much easier than a 4th-order one for the above specifications.

2. PROTOTYPE SYSTEM-LEVEL DESIGN

2.1 Multi-notch FIR Transfer Function

Both FIR and IIR functions can be chosen for the design of this bandpass interpolating filter. It needs a 6th-order IIR function with a Q-factor higher than 20. However, it is not possible to obtain accurate passband location by adjusting the sampling rate, which can be normally used for standard high-Q SC filtering [6.19, 6.20], as the frequency-translated band will also be shifted according to the change of sampling rate. Besides, the required non-recursive filter taps after IIR multirate transformation for ensuring low-speed operation of circuit [6.21] will still be as high as 25 terms. An FIR function with its low sensitivity nature in the passband is more suitable for this application. Nevertheless, for traditional designs, the filter taps can be as high as about 30 based on Optimum Parks-McClellan or linear programming algorithms, and more than 40 for the normal Windowing method. The coefficient spread is also unacceptable in those approaches. In order to overcome such design difficulties, a specific Optimum Zero-Placement methodology is employed here, as shown in the z-domain zero-plot of Figure 6-2, whereby some zeros are placed directly at the input and unwanted imaging bands, e.g. two zeros inside both 22-24 MHz and 102-104 MHz bands; some are located nearby unwanted imaging bands for helping the rejection with other extra functionality, e.g. zero at 80 MHz is specifically added for eliminating DC modulation to the DAC DC offset that is associated with the filter input signal, thus improving the output signal purity, as the resulting spurs are near passband but without enough attenuation. The final zero location and number are optimized according to the required rejection, together with a Monte-Carlo sensitivity analysis to take into account the practical capacitance ratio mismatches. The filter finally requires only 15 taps in Table 6-1 and yields a very satisfactory maximum spread of 6. For having stopband rejection around 45 dB, the standard deviation of capacitance mismatches must be controlled within 0.4 %, regardless of all other factors.

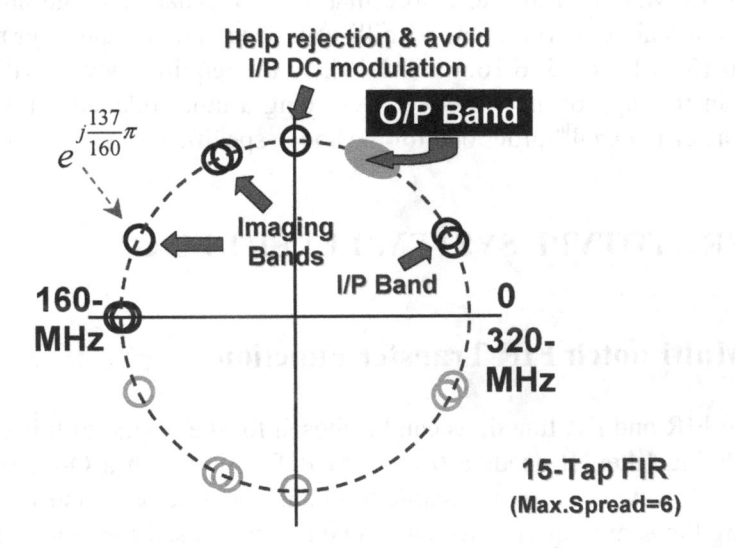

Figure 6-2. Zero-plot for multi-notch FIR system function by optimum zero-placement method

Table 6-1. Tap-weight for multi-notch FIR system function

Tap-weight	Value
$h_0 \mid h_{14}$	0.2503
$h_1 \mid h_{13}$	0.4830
$h_2 \mid h_{12}$	0.3152
$h_3 \mid h_{11}$	-0.1878
$h_4 \mid h_{10}$	-0.5758
$h_5 \mid h_9$	-0.1708
$h_4 \mid h_8$	0.6346
h_7	1

2.2 Time-Interleaved Serial ADB Polyphase Structure with Autozeroing

Since the filter length is greater than two times of the interpolation factor 4, the canonic ADB polyphase structure is suitable for circuit implementation with high-frequency operation as mentioned in Chapter 2 / Session 4. According to (3.17), for ensuring 45 dB signal to pattern noise ratio, the channel offset mismatches must be controlled with the standard deviation smaller than 2 mV which is very hard to achieve for this filter length and operating frequency. Therefore, autozeroing technique is employed finally for overcoming these errors. However, to cope with the achievable delay for the autozeroing SC delay cells which have one output phase for resetting, and also to minimize the number of power-consumed & error-accumulated serial ADB's, a specific time-interleaved serial ADB polyphase structure with autozeroing configuration is designed here, as shown in Figure 6-3. This structure realizes the required delay for coefficients by an optimal combination of time-interleaving delay path with multiple unit delay cells, delay-free or multi-unit delay SC polyphase branches for positive and negative coefficients, as well as a special multiplexing network. Finally only 4 ADB's are required for this 15-tap transfer function. The 4 individual accumulators for each polyphase subfilters, which are also the most power-consumed and accuracy-dependent part of the filter core, fully operate at input lower 80 MHz, and thus leaving a relatively-simple-operation output multiplexer operating at higher output rate that won't cost too much as will be discussed next.

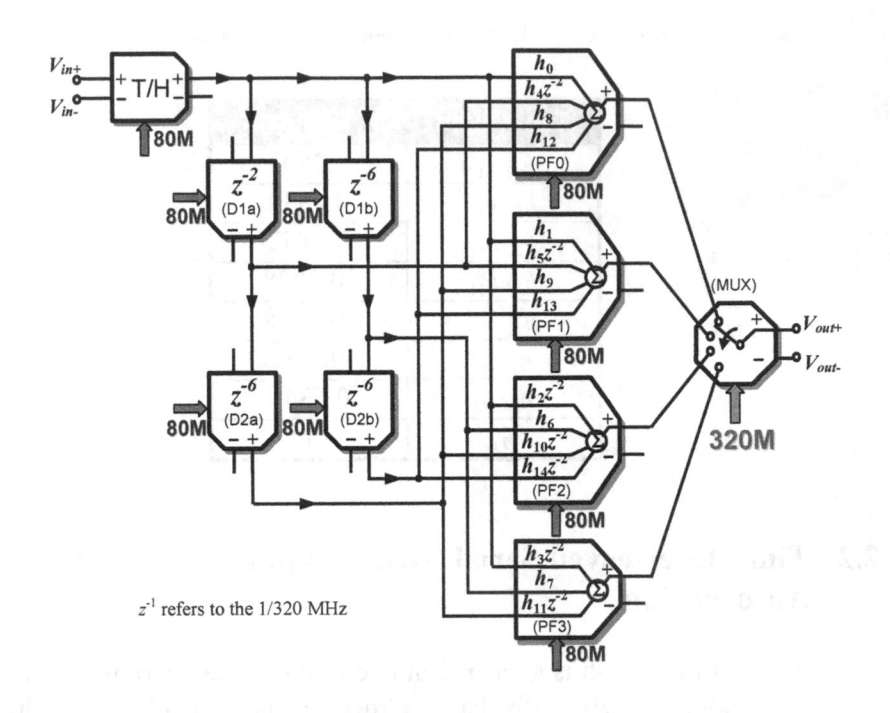

Figure 6-3. Time-interleaved serial ADB polyphase structure with autozeroing

3.   PROTOTYPE CIRCUIT-LEVEL DESIGN

3.1   Autozeroing ADB and Accumulator

The circuit is implemented in a fully-differential architecture and the autozeroing SC ADB is the modified symmetrical version of the mismatch-free autozeroing SC delay circuit presented in Chapter 4 / Session 3. For simplicity, only the SC ADB (D1b) with multiunit delay z^{-6} (or $z^{-3/2}$ referred to input lower sampling rate) is presented in Figure 6-4. Note that the extra switches *sw1* & *sw2* (*sw1'* & *sw2'*) are mandatory for breaking the resistive paths during the charge-holding phase of SC branch *1* and *2*, thus eliminating the signal-dependent clock-feedthrough and charge-injection errors. Since both the slew rate and feedback factor of the close-loop delay circuit are proportional to the capacitance value of sampling and holding capacitor, a value of compromise is found at 0.48 pF, which has also enough noise headroom for the required noise tolerance of the filter.

The fully-differential autozeroing SC accumulators are illustrated in Figure 6-5(a) and (b) for polyphase subfilter $m=0$ and $m=2$, respectively. The filter tap weights are implemented as direct capacitance ratios of the SC branches to the summing capacitor by either in-phase direct charge coupling or out-phase charge transferring. Especially, to help the minimization of the time-interleaved paths for the last high-speed multiplexer with also the full output sampling period for settling, the polyphase subfilters $m=0$, 1 and $m=2$, 3 are designed to operate at different A and B phases, respectively, thus having 2 extra unit delay embedded in all coefficient branches for subfilters $m=2$, 3. This special arrangement must also be taken into account for the realizations of the coefficient delay, as well as the structural optimization in terms of number of ADB's that have physical reset phase existence for autozeroing.

Due to noise, matching and speed considerations, the unit capacitance is nearly 100 fF and 150 fF, respectively for polyphase subfilters $m=0$, 1 and $m=2$, 3, thus yielding the final adjusted maximum spread of 8, as shown in the Table 6-2. Instead of normalizing the smallest weights, which normally has less matching sensitivity, to the unit capacitance, the summing capacitors are normalized to the nearest integer multiples to the unit capacitance so as to eliminate the need for non-unit capacitor for further improvement of the ratio matching.

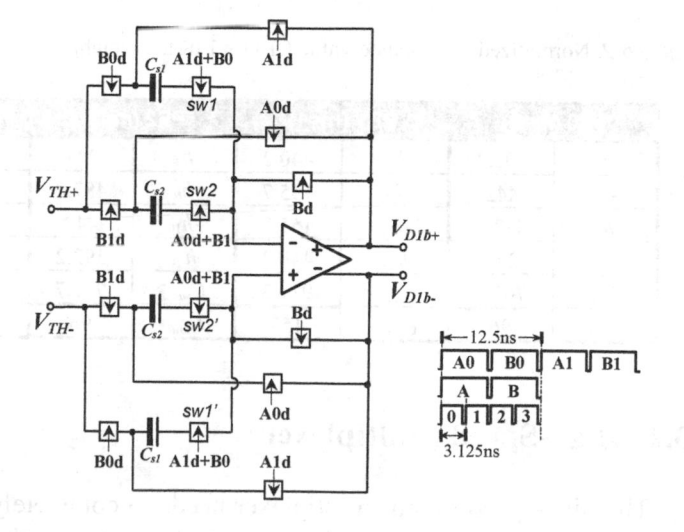

Figure 6-4. Autozeroing, Mismatch-Free SC ADB with z^{-6} delay

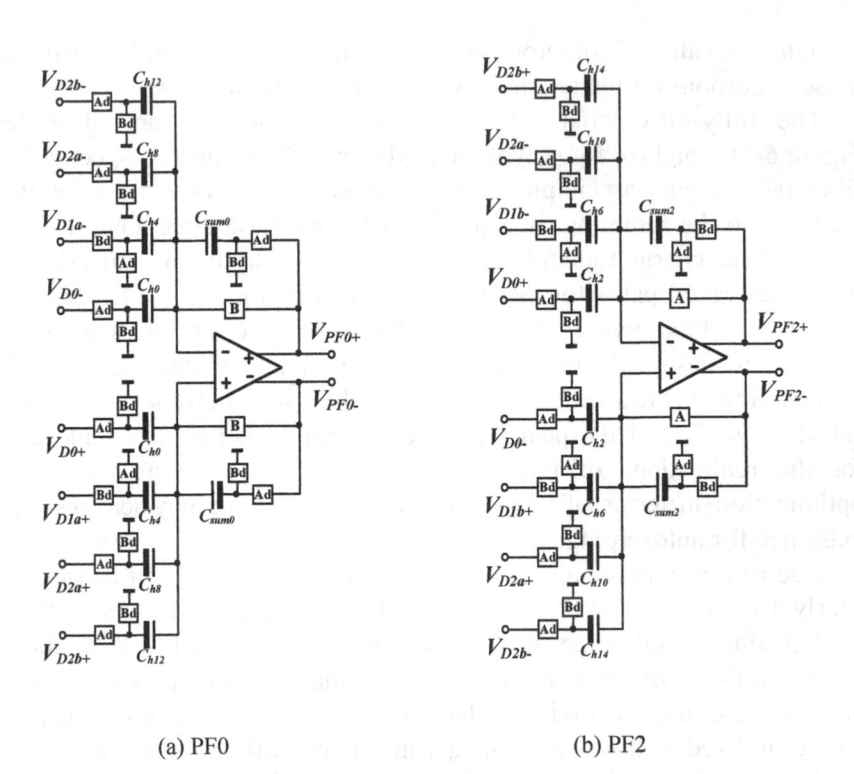

(a) PF0   (b) PF2

Figure 6-5. Autozeroing SC accumulator for polyphase subfilter (a) *m*=0 (b) *m*=2

Table 6-2. Normalized capacitance value (fF) for FIR tap-weight

PF0 (m=0)		PF1 (m=1)		PF2 (m=2)		PF3 (m=3)	
h_0	192.2	h_1	440.2	h_2	242	$-h_3$	171.1
$-h_4$	442	$-h_5$	155.7	h_6	487.2	h_7	911.3
h_8	487.2	$-h_9$	155.7	$-h_{10}$	442	$-h_{11}$	171.1
h_{12}	242	h_{13}	440.2	h_{14}	192.2		
C_{sum0}	767.7	C_{sum1}	911.3	C_{sum2}	767.7	C_{sum3}	911.3
C_{unit0}	96	C_{unit1}	152	C_{unit2}	96	C_{unit3}	152

3.2 High-Speed Multiplexer

The high-speed output multiplexer needs to completely settle within 2.4 ns for providing 320 Msample/s outputs by counting the non-overlapping phase gap. Without the employment of autozeroing, the circuit, as shown in

Figure 6-6, is optimized to have full output sampling period operation at maximum speed, while minimizing power consumption and improving linearity, because the input referred DC-offset of multiplexer here will result mainly in a DC level shifting rather than the pattern noise tones. The charge-transferring free property boosts maximally the operating speed by the enlarged feedback factor and it reduces further the path gain and bandwidth mismatches associated with double-sampled SC Common-Mode Feedback (CMFB). Note that the sampling/holding capacitor mismatching still renders the speed mismatches among 4 paths, which, however, can be neglected, e.g. 5% capacitor mismatch results in the modulation sidebands always below -60 dB. The noise and gain errors are low by using sampling/holding capacitor of 0.7 pF, thus yielding an efficient trade-off between feedback factor and slew rate. Especially, the charge-transferring free property also reduces the output glitch effects that normally happen in SC circuits due to the charge redistribution using high-output-impedance transconductance opamp.

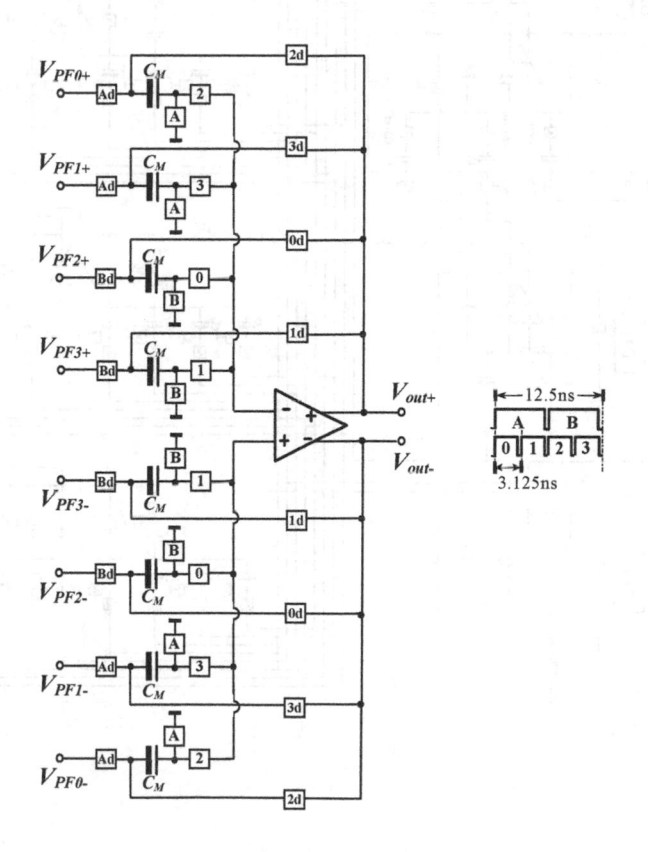

Figure 6-6. High-speed mismatch-free SC multiplexer

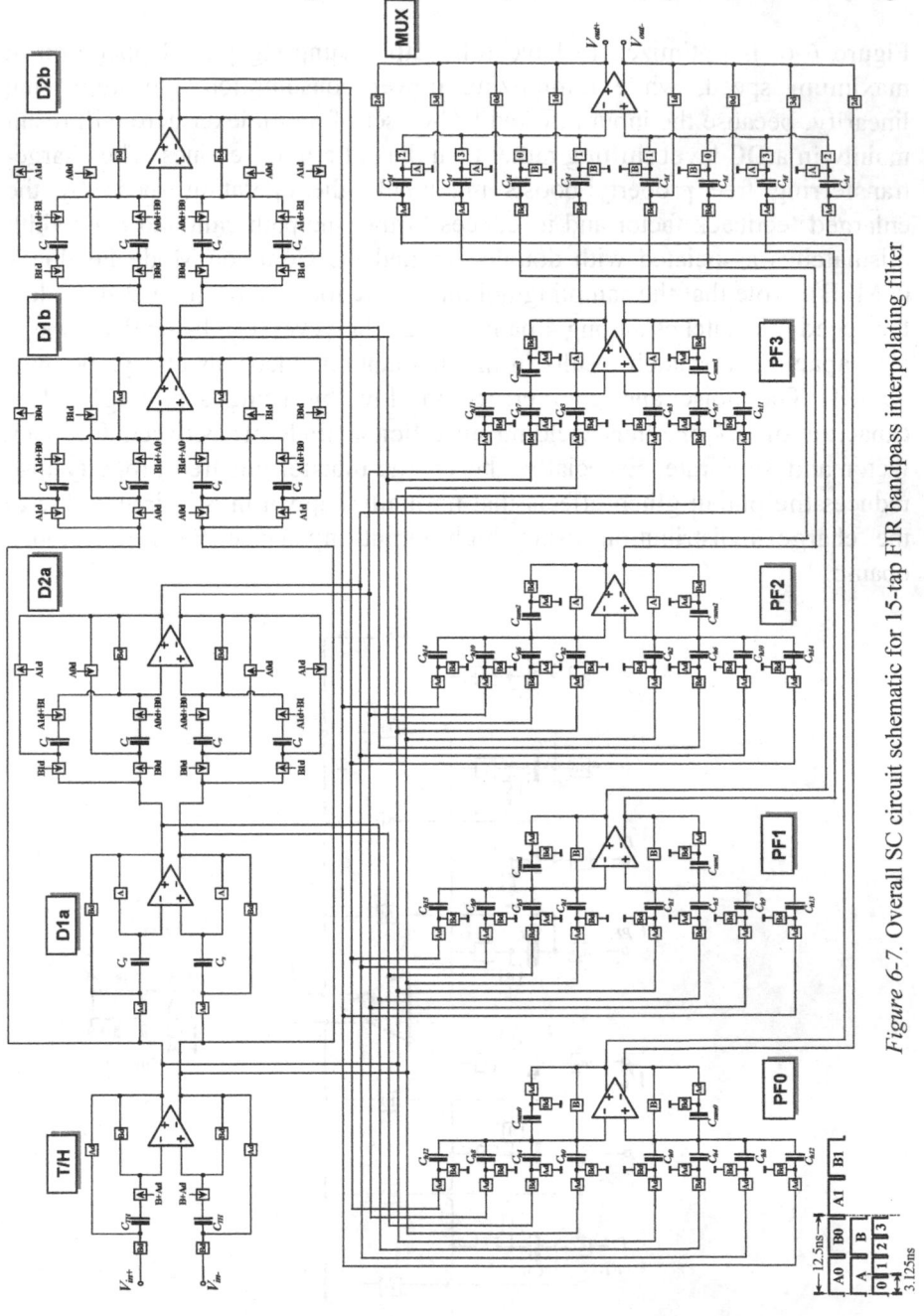

Figure 6-7. Overall SC circuit schematic for 15-tap FIR bandpass interpolating filter

3.3   Overall SC Circuit Architecture

The complete fully-differential SC circuit schematic is presented in Figure 6-7 with simplified clock phases. The circuit is implemented in 0.35 µm double-poly, triple-metal CMOS technology with a single 2.5 V supply and 0.95 V common-mode level. The complete system contains an input testing-purpose autozeroing Track-and-Hold (T/H), time-interleaved serial delay line composed by autozeroing mismatch-free ADB's, 4 path polyphase subfilters with autozeroing SC accumulators, as well as the output high-speed mismatch-free SC multiplexer. Besides, the real clock phases possess also the falling-edge delayed one for the bottom-plate sampling so as to reduce the signal-dependent charge-injection and clock-feedthrough effects.

3.4   Telescopic opamp with Wide-Swing Biasing

For this design, less than 9.5 ns and 2.2 ns settling times are required for the opamps in low-speed filter core and high-speed multiplexer, respectively, but the minimum gain needed ranges from 1000-1500 without big ill effects from the behavior model verifications. Single-stage transconductance opamp with cascode active loading is especially appropriate for the high-speed but moderate gain requirements, as the non-dominant pole locates normally far from the dominant one. Telescopic cascode topology is employed here for its superior high-speed (only nMOS transistors vs. both nMOS and pMOS for folded-cascode in signal path) and low-power (2 vs. 4 current legs), as well as low-noise capability (4 vs. 6 noise contributing devices) when compared with the folded-cascode opamp, but with the price of having one less effective voltage in the output swing. The complete telescopic opamp structure with it biasing circuitry is presented in Figure 6-8.

The nMOS differential pair M_1 and M_2 is used to achieve a larger transconductance with reduced input capacitance. Moreover, capacitive neutralization technique by the cross coupled, drain-source connected transistors M_{m1} and M_{m2} are used to cancel the Miller multiplication effects that will result in large input capacitance and thus reduce the gain accuracy as well as the operation speed. Letting the M_{m1} and M_{m2} half size of input devices M_1 and M_2 to make these auxiliary capacitors approximately equal to the gate-drain capacitance of the input devices, lead to the miller multiplication factor reduced from $(1+g_{m1}r_{ds1})$ to 2 only. Note that the junction capacitance associated with M_{m1} and M_{m2} increases the capacitance at the source of the cascode device, which contributes directly to a decrease in the location of the nondominant pole, thus reducing the phase margins.

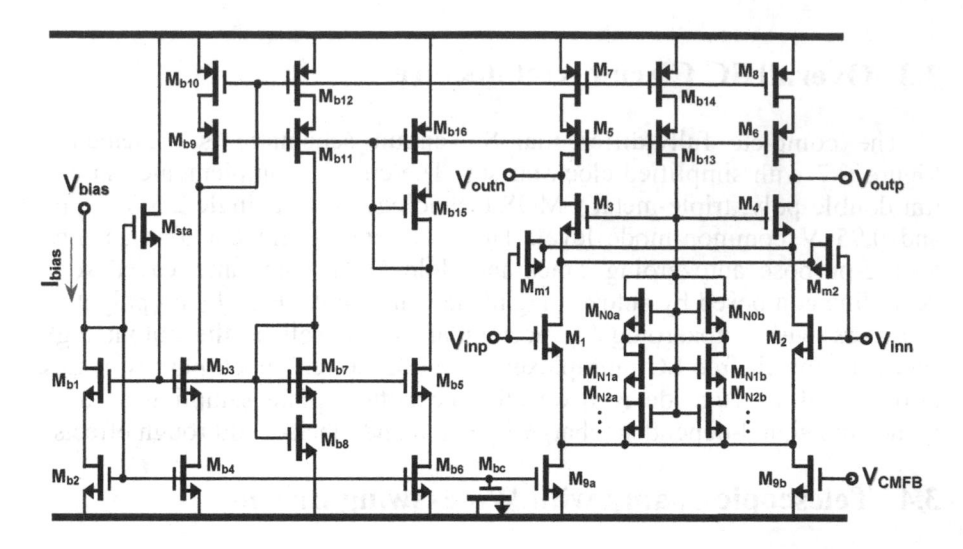

Figure 6-8. Schematic of Telescopic opamp with wide-swing biasing

Table 6-3. Device size for Telescopic opamp and wide-swing biasing circuitry

Transistor	Size (μm)	Transistor	Size (μm)
M_1, M_2	15.6/0.35×32	M_{m1}, M_{m2}	15.6/0.35×16
M_3, M_4	17.5/0.5×40	$M_{N0a,b}, M_{N1a,b}$ ...	17.5/0.5
M_5, M_6	12.5/0.7×40	$M_{b9}, M_{b11}, M_{b13}, M_{b15}$	12.5/0.7×2
M_7, M_8	23/1.15×40	$M_{b10}, M_{b12}, M_{b14}$	23/1.15×2
M_{9a}	20/0.7×48	$M_{b1}, M_{b2}, ... M_{b7}$	20/0.7×2
M_{9b}	20/0.7×32	M_{b8}	20/3.5×2
M_{sta}	2/0.3×2	M_{b16}	12.5/4.3×2

Without the loss of the required output swing, a proper wide-swing biasing circuitry is designed so as to make the input and output common-mode levels of the opamp both equally at 0.95 V due to the reset nature of autozeroing techniques used in the filter core. All the transistors in the opamp are properly designed with the gate-source effective voltage $V_{eff} = V_{GS}-V_{th}$, around 100-150 mV for nMOS and 200 mV for pMOS, and especially, they are all biased such that their drain-source voltage, V_{DS}, is greater than their effective voltage by more than 100 mV to insure that the transistors remain well within the saturation region. Hence, the 1.2 V_{p-p} output voltage swings limited mainly by the lower swing bound can be normally guaranteed.

All the cascode devices are biased by gate-connected stacked devices, and the top (for nMOS cascode) / bottom (for pMOS) one is diode-connected in the saturation region, e.g. M_{b7}, M_{b15}, and another is in the triode region, e.g. M_{b8}, M_{b16}, which emulates the voltage drop V_{DS} across the current source transistor below cascode device. Especially, for proper biasing the nMOS cascode devices M_3 and M_4, instead of a single triode device with longer channel length, several stack triode devices, $M_{N1a,b}$, $M_{N2a,b}\cdots$, with the same channel length as M_3 and M_4 are used in order to have better matching. This is because that the biasing of the nMOS cascode devices are the most dominant and sensitive part which is related to the proper output swing and linearity especially when opamp is designed with the same input and output common-mode levels at a low value. Furthermore, this biasing for nMOS cascode devices is so-called internal biasing, as the bias voltage is referred to the center point of the source-coupled pair instead of ground so that it tracks the movement of the input common-mode, thus alleviating sensitivity to the process variation. The small transistor M_{sta} insures that the biasing circuit settles to the desired operating point instead of a stable, zero-current state.

The proper sizing of the opamp transistors is also necessary for acquiring better matching according with layout considerations, e.g. the biasing transistors are always configured in two unit parallel forms for meeting the layout symmetry, and after counting the current mirror ratio 20 of the main opamp core and the biasing, the opamp transistor must be 40 times higher than the biasing unit transistor so as to keep the same unit transistor for better matching to avoid the Dog-bone structure [6.22] in current mirrors. Thus, the opamp is designed with respect to the sizing to the unit transistor. The final transistor sizing for the fastest opamp used in the circuit is shown in the Table 6-3.

To provide good isolation among the opamps on the chip, each opamp has its own local biasing circuit that is activated by a master bias current distribution network. This master bias current is generated on chip by mirroring a current from an external current sink, and distribute to local biasing circuit as a current through pMOS cascode current mirroring so that the voltage drops in routing do not affect the biasing. Especially, in order not to waste the unused layout area around the opamp, the drain-source connected nMOS capacitor (M_{bc}) is added at the bias line of the tail current source so as to improve the Common-Mode Rejection Ratio (CMRR) and Power Supply Rejection Ratio (PSRR).

To define proper DC level at the high-impedance output nodes of opamp, the dynamic SC Common-Mode Feedback (CMFB) circuit is employed for attaining larger output swing but without dissipating DC static power. As shown in Figure 6-9, the single-sampling CMFB is used for the opamps in autozeroing filter core where the C_1 and C_2 are 90 fF and 200 fF, respectively; while a double-sampling version, as shown in Figure 6-9(b), is specially designed for the high-speed output multiplexer for better linearity and channel matching.

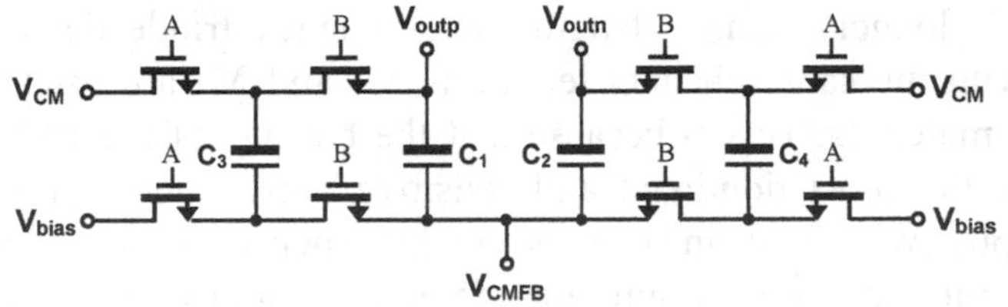

(a)

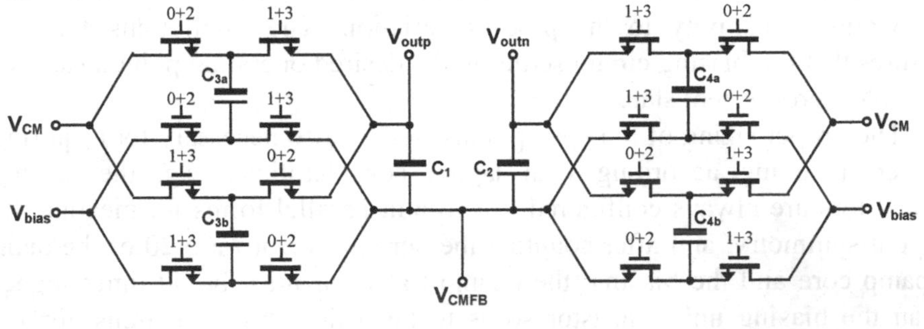

(b)

Figure 6-9. (a) Single-sampling SC CMFB for filter core and (b) Double-sampling SC CMFB for multiplexer

3.5 nMOS Switches

The circuit analog common-mode level is set at a low level of 0.95 V so that only nMOS switches can be used to minimize clock distribution routing and synchronization complications due to the multirate/multi-phase scheme, as well as the digital noise coupling in high-frequency operation. The on-resistance is required to be low enough over the complete range of the input signal swing so as to alleviate its non-linear property and have complete RC settling, as well as reduce the thermal noise contributions mainly for the switches in the signal or charge transferring path during the output phase. Furthermore, high on-resistance in nMOS switches especially located in the feedback path can slow down the circuit and also make the feedback system poorly damped or even unstable. This results from the increase in phase shift by increasing the delay and thereby reducing the phase margin. However, using too large switch not only adds significant amount of drain/source junction parasitic capacitance at the output but also, more importantly, leads to increased charge-injection and clock feedthough errors. All the switches are sized distinctly to accommodate different capacitor loading according to the rule-of-thumb that requiring RC bandwidth to be greater than four times the operating sampling rate in the worst corner case. The final obtained switch size ranges from the maximum at 53 μm /0.3 μm to the minimum at 5 μm /0.3 μm.

3.6 Noise Calculation

Due to the employment of autozeroing technique, the thermal noise from switches and opamps is the dominant noise contribution for the total noise, as the flicker or 1/f noise of opamp will be greatly attenuated. The estimated total output noise is calculated according to the noise analysis procedure in Chapter 3 / Session 7 as shown in the Table 6-4, where the parameters like on-resistance and transconductance are derived from the transistor simulations, and the excess noise factor in spectrum density of telescopic opamp is

$$\gamma = 1 + \frac{g_{m7}}{g_{m1}} \qquad (7.1)$$

which is normally smaller than 1.5, thus implying that the telescopic opamp has the lowest noise among all other cascode single-stage opamp structures.

Table 6-4. Noise contributions

Noise Contribution	ADB Delay Line	Polyphase Subfilter	MUX	Total Output Noise
$4f_s = 160$ MHz	$48.9\ nV_{rms}^2$	$9.6\ nV_{rms}^2$	$8.8\ nV_{rms}^2$	$67\ nV_{rms}^2 / 259\ \mu V_{rms}$
$4f_s = 320$ MHz	$53..8\ nV_{rms}^2$	$10.3\ nV_{rms}^2$	$9.1\ nV_{rms}^2$	$73\ nV_{rms}^2 / 271\ \mu V_{rms}$
$4f_s = 400$ MHz [1]	$49.8\ nV_{rms}^2$	$9.7\ nV_{rms}^2$	$7.9\ nV_{rms}^2$	$67\ nV_{rms}^2 / 260\ \mu V_{rms}$

[1]: DVDD = 3.3 V, this is for comparison purpose to the measurements that will be discussed in next chapter

3.7 I/O Circuitry

To emulate the actual sampled-and-held signals generated from DDFS logic and DAC, an input T/H SC stage is designed with autozeroing and mismatch-free properties for sampling signals at lower 80 MHz for testing purposes only. This circuit, as shown in Figure 6-7, has similar architecture to the ADB cell with a 0.6 pF sampling capacitor, and the additional switches between the sampling capacitor and opamp input are specially used to eliminate the undesired glitch noise due to the capacitive direct coupling from input for a clean signal environment.

An output driver including a level-shifter followed by a low-output impedance buffer are also employed next to the output PAD in order to drive 50 ohms input-impedance equipment for such high frequency testing. Due to the requirement of high current driving capability with together high linearity and bandwidth, the simple source-follower structures are employed for the output driver. For an accurate evaluation to the S/H characteristic of the output signals at 320 Ms/s, they are designed to have a very wide, more than 1 GHz, -3dB bandwidth and <-65 dB THD in simulations.

3.8 Low Timing-Skew Clock Generation

As discussed in Chapter 3, the timing-skew effects due to the input sampling of 4 parallel polyphase filter bands are negligible due to the S/H input signal nature, and thus the timing-skew errors are mainly imposed by the last high-speed multiplexer stage for switching among 4 parallel subfilters at 320 MHz. Different from input sampling where the error tones can be shaped by system function, here the resulting modulated sidebands will fold back inside the stopband that cannot be removed by the filter. The Figure 6-10 shows the means of the SNR and SFDR for the circuit operating at 160, 320 and 400 MHz sampling frequencies from 100-time Monte-Carlo

MATLAB modeling simulation results. It is obvious that it requires σ<5 ps
to ensure the mean of SFDR>60 dBc for 320 MHz sampling rate in order not
to degrade the overall system response.

To reduce the signal-dependent clock-feedthrough and charge-injection
errors, the clock-delayed (bottom-plate sampling) technique is employed,
and thus, totally 21 phases are needed where 8 and 13 phases rotating at 320
and 80 MHz, respectively, from a master 320 MHz reference. The simplified
structure for this multi-phase generation is presented in the Figure 6-11.

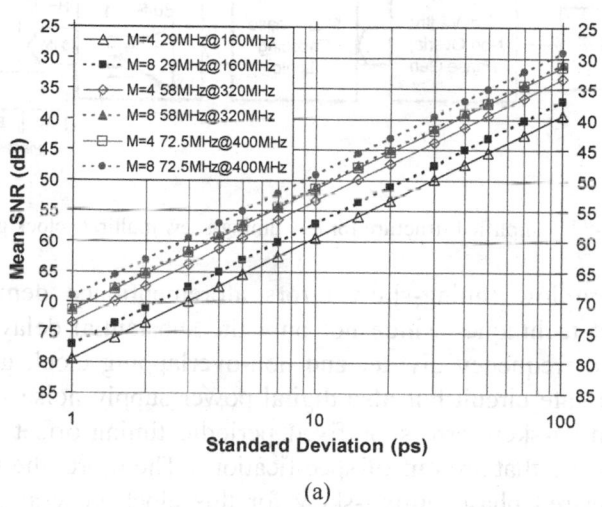

(a)

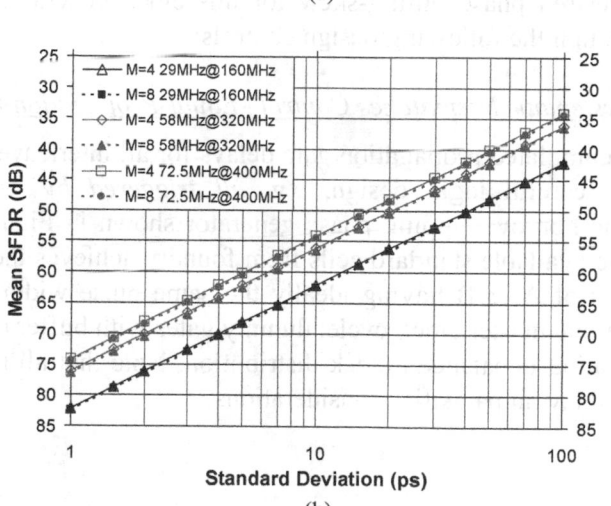

(b)

Figure 6-10. (a) SNR and (b) SFDR Mean vs. timing-skew errors and sampling rates (100-
time Monte-Carlo)

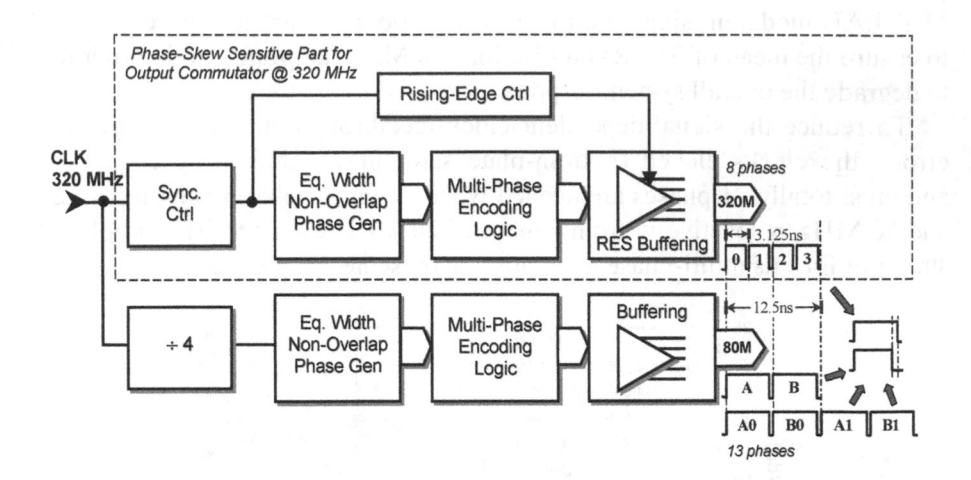

Figure 6-11. Simplified structure for low timing-skew multirate clock generator

The severe low timing-skew errors aforementioned demand specific digital circuit techniques, since not only the substantial delay mismatches from standard frequency divider and non-overlapping clock generation, as well as logic gate circuit but also digital power supply noise will all easily introduce timing-skew errors or fixed periodic timing-offset among time-interleave phases that are out of specifications. Therefore, the minimization of such undesired phase timing-skew for this clock generator is specially achieved through the following design controls:

I. *Design Systematic Mismatches Control – Equal-Propagation-Gate- Delay:*

The accumulated propagation gate delays for all interleaved phases are balanced by careful logic design, e.g. *all triggered by same edge of reference*: the non-overlapping phase generator shown in Figure 6-12 with the use of the available standard cells from foundry achieves the rising (also falling) edges of A & B having ideally the same pulse width with timing difference of one master clock cycle; dummy gates with buffer trees are used for compensating unbalanced clock distribution. Note that all those designs have included layout parasitics considerations.

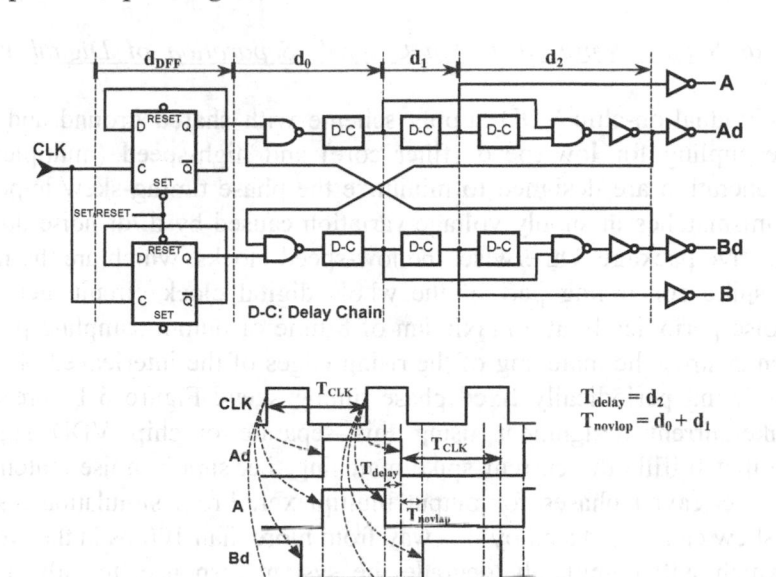

Figure 6-12. Equal-width non-overlapping clock phase generation

II. *Random Process Mismatches Control – Output Rising-edge Synchronization:*

To minimize the random process mismatches imposed from the logic gates in skew-insensitive time-interlcaved phase generation path, a specific Rising-Edge Synchronizing (RES) buffer array triggered by a single clock is designed as shown in Figure 6-13. It locates just before the last buffers that drive the clock buses for generating 8 phases for the output high-speed multiplexer. This implies that the random mismatch ideally happens only in the last large buffer that can be indeed neglected.

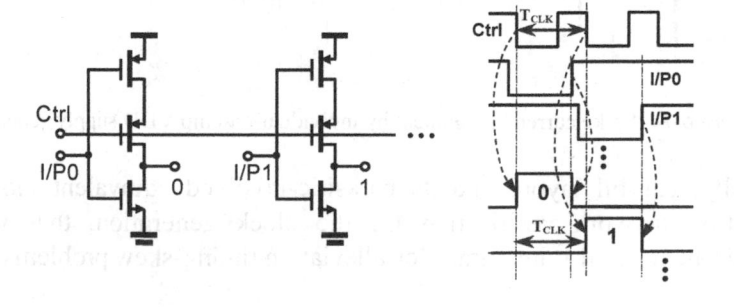

Figure 6-13. Rising-edge-synchronization buffer array

III.   dI/dt Supply Noise Mismatch Control: Separation of Digital VDD
Supplies

Individual-on-chip-VDD supply scheme with shared ground and on-chip decoupling for low-speed (filter core) and high-speed (multiplexer) clock generation are designed to minimize the phase timing-skew imposed by the mismatches in supply voltage variation caused by dI/dt noise due to the inductive package. Otherwise, the low-speed clocks, which are the most current-spike consuming part of the whole digital clock circuit, generate dI/dt noise periodically at a maximum of 8-time of output sampling period and then destroy the matching of the rising edges of the interleaved phases, thus rendering periodically fixed phase timing skew. Figure 6-14 presents the spike-current assignment using this separate on-chip VDD supply scheme that fulfills the current spike matching, i.e. supply noise matching, among interleaved phases for output multiplexer. From simulations, such timing skew can be reduced by this way from more than 100 ps in the worst-case, which will completely degrade the system response, to only about several pico-second level.

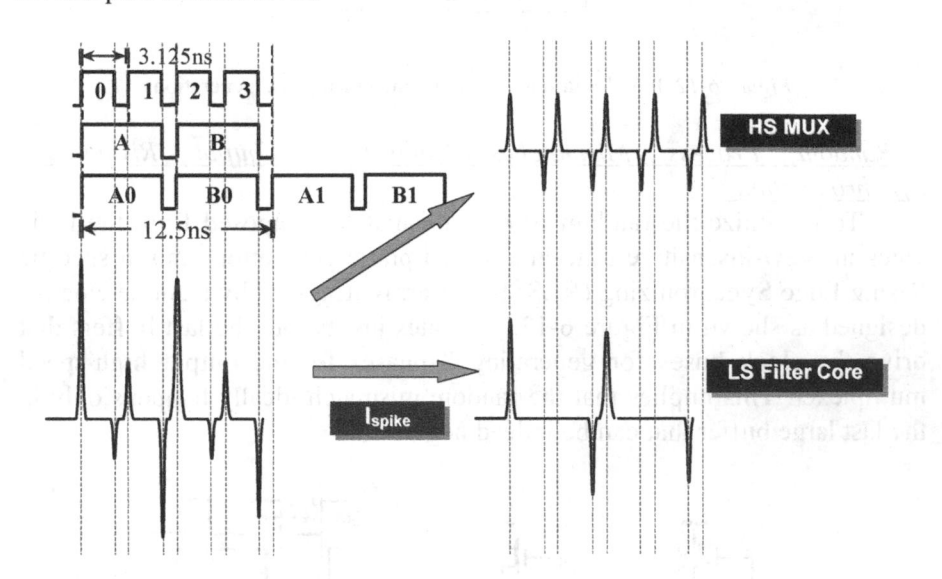

Figure 6-14. Spike current assignment by individual-on-chip VDD supply scheme

Finally, careful layout and its parasitic-involved equivalent clock bus distribution network verification for the clock generation, that will be discussed next, is also important for alleviating timing-skew problems [6.23].

4. LAYOUT CONSIDERATIONS

This high-speed bandpass SC interpolating filter prototype is laid out in 0.35 μm double-poly, triple-metal CMOS with lightly-doped substrate. The practical layout issues such as: device matching and routing of critical paths, parasitic effects, power and ground isolation, noise coupling from digital section to the analog, and etc, which have both significant impact for the overall performance especially for this very high-frequency operation, are underscored in the following techniques

4.1 Device and Path Matching

To deal with the random process variation, the matching is achieved by concentrating on each building block especially any parallel part from top-to-bottom manner, i.e. polyphase subfilters, ADB's, clock generation to opamp and capacitor group, etc. The strict mirror symmetry or common-centroid with dummy periphery techniques [6.24, 6.25, 6.26, 6.27, 6.28, 6.29, 6.30] are applied for those parts.

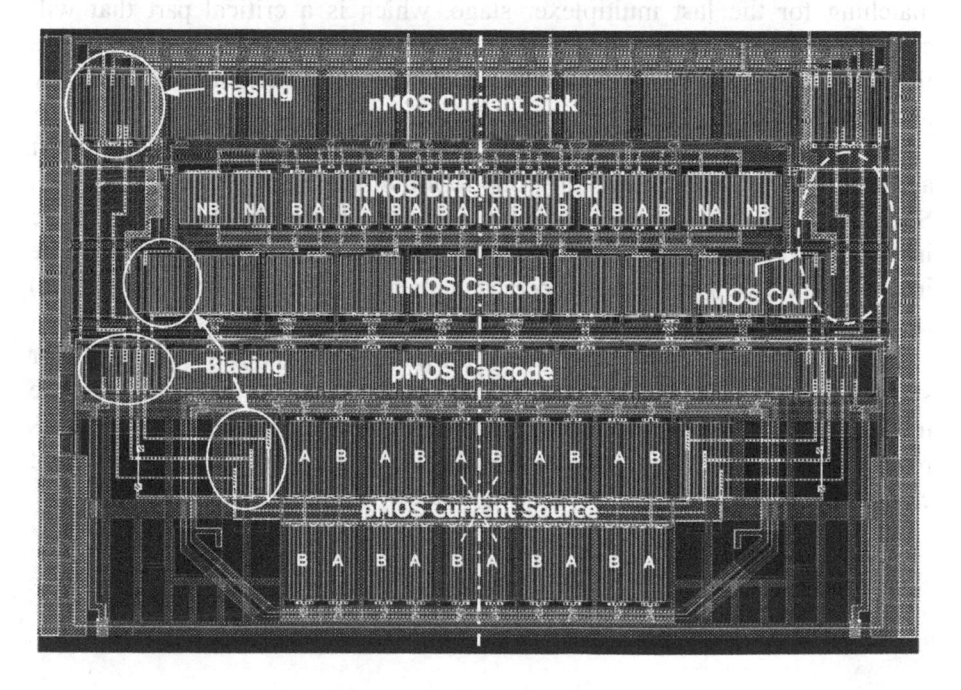

Figure 6-15. Layout of Telescopic op amp

Although the filter core is offset-compensated, better opamp matching will improve significantly the CMRR and PSRR that is also vital for mixed-signal high frequency operation. As presented in opamp layout of Figure 6-15, only the matching-sensitive nMOS differential pair and pMOS current source are laid out with common-centroid technique. The whole opamp including biasing circuitries, which are located at the left and the right sides, are exactly mirrored symmetrically around the centerline. The unused places are also filled with nMOS capacitor at tail-current source biasing line.

Capacitors for matching-sensitive coefficients are laid out carefully using common-centroid unit-capacitor array surrounded by non-unit capacitors for alleviating the oxide-thickness gradient effect, as shown in chip microphotograph of Figure 6-16(a). The capacitor ratio errors due to overetching are minimized by keeping perimeter-to-area ratios the same for all unit and non-unit capacitors. Full-unit capacitor arrays are used for the edge dummy to keep the similar environment more than 20 μm range to the capacitor group for minimizing the proximity effects [6.31], and they are also specially employed for the analog supply decoupling. The capacitors for SC ADB's are simply laid out with single unit due to its mismatch-free circuit nature, as shown in Figure 6-16(b). To further minimize the path matching for the last multiplexer stage, which is a critical part that will contribute for the mismatch errors to the output directly, the capacitor are formed by 4 unit capacitors with strict isolation to other path capacitors by 3-side shielding, as shown in Figure 6-16(c).

The capacitors for dynamic SC CMFB are located symmetrically at left and right side of each capacitor group to minimize the digital clock routing. Special cares are also taken during the routing of the top and bottom plate interconnection lines of the capacitor groups in order to minimize the internal and external interconnect capacitance (fringing and overlapping) mismatches [6.31].

Each polyphase filter and ADB blocks are laid out in strict mirror symmetry as shown in Figure 6-17 for a polyphase subfilter $m=0$. All the metal line width in the layout is determined carefully by the allowed series resistance as well as the peak current density associated with the sufficient contacts and vias in order also to minimize the resulting thermal noise.

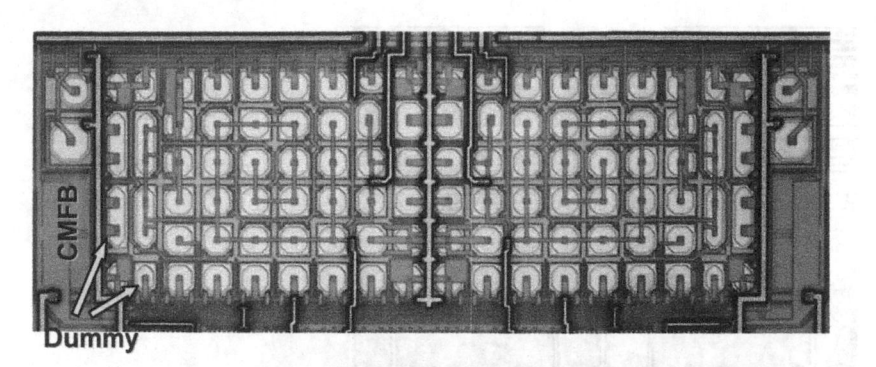

(a) Capacitor group for polyphase subfilter $m=0$

(b) Capacitor group for z-6 ADB

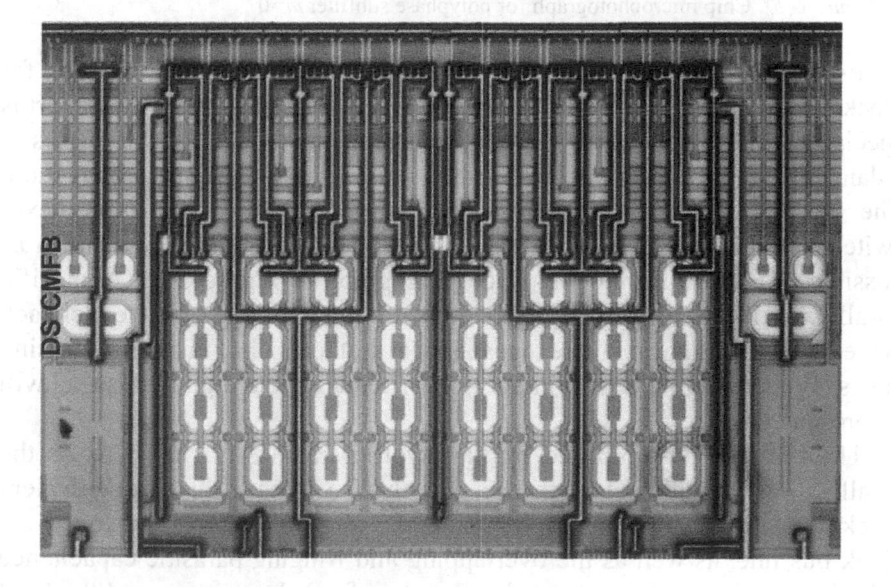

(c) Capacitor group for multiplexer

Figure 6-16. Chip microphotograph for capacitor group for (a) Polyphase subfilter $m=0$ (b) z^{-6} ADB   (c) Multiplexer

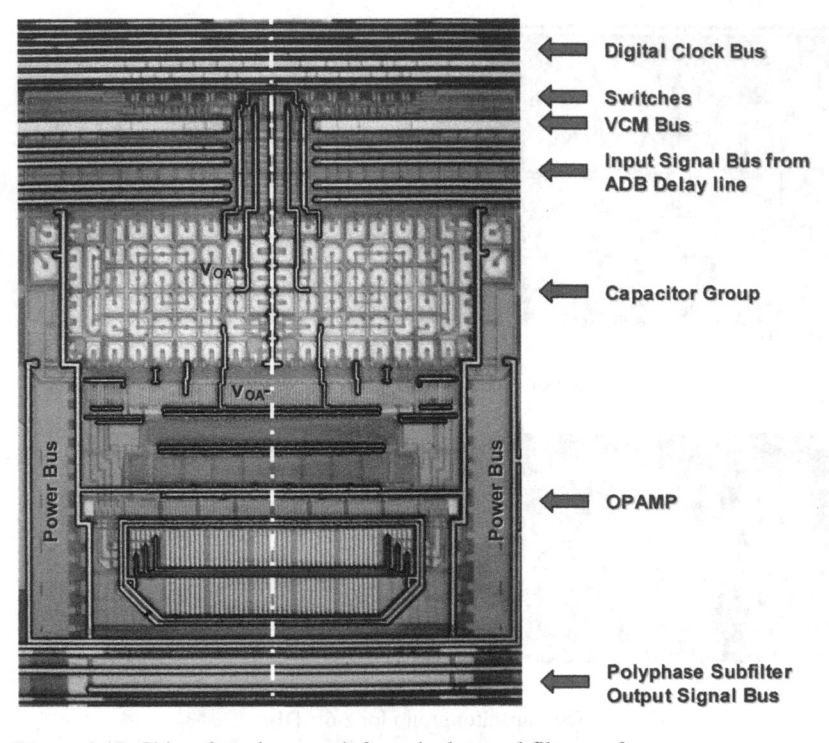

Figure 6-17. Chip microphotograph for polyphase subfilter $m=0$

Especially, due to the severe timing-skew tolerance, all digital cells for clock generation are customly laid out. The sensitive high-speed clock part is specially laid out with mirror parallel arrangements for interleaved phases to balance the systematic parasitics loading associated with signal propagation. The clock bus routing from the clock buffer to the output multiplexer switches is minimized so as to reduce the random mismatches, as much as possible, as shown in the chip microphotograph of Figure 6-18. A specific parallel clock-line routing arrangement is designed to minimize the kickback noise from neighboring wire capacitance that can also disturb the rising edges. Well shielding is also important for reducing the digital noise, as will be presented next.

The timing delay imposed by the layout parasitics is vital here, as the smallest pulse width is less than 3 ns only. To foresee this, an equivalent clock bus distribution network containing transmission resistance of the clock bus line, as well as the overlapping and fringing parasitic capacitance is built according to the extracted parameters from the process, and the final layout is subject to the simulation results for ensuring such timing errors don't have harmful effects into the clock phases.

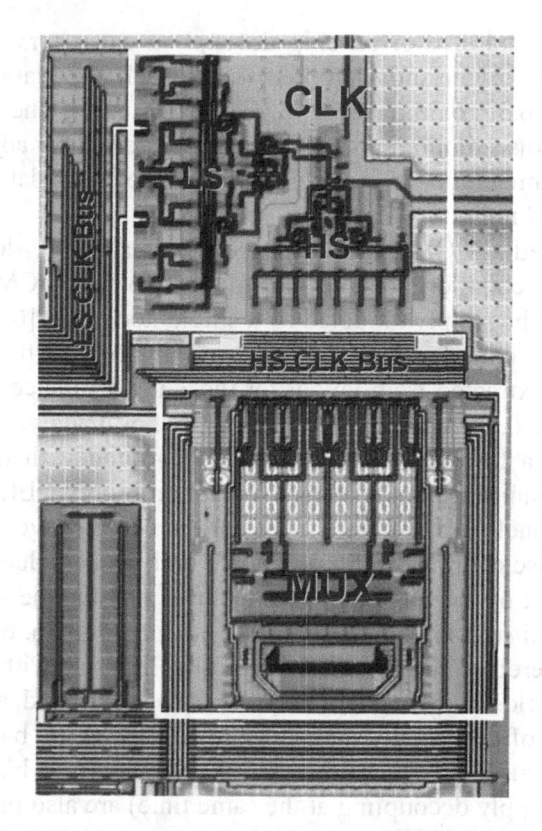

Figure 6-18. Chip microphotograph for clock generator and output multiplexer

4.2 Substrate and Supply Noise Decoupling

All current injected into the common substrate results into fluctuations of the substrate voltage, namely substrate coupling, which is identified as a major problem in mixed-signal IC's. It is mainly caused by 3 mechanisms: coupling from noisy digital supply, capacitive coupling (e.g. NWELL, junction capacitor, node parasitics, etc.), as well as impact ionization in the MOSFET channel [6.32, 6.33, 6.34, 6.35, 6.36, 6.37, 6.38]. The digital supply noise imposed by the dI/dt noise due to the inductive package is often the dominant contributor for substrate noise coupling [6.36], and at the same time, it also disturbs the quiet analog supply domain. Therefore, the common rule-of-thumb for isolating their direct coupling is to simply separate the analog AVDD and digital DVDD pins [6.32, 6.33, 6.34, 6.39, 6.40]. However, ground pins and the bulk-biasing alternative is highly dependent on the substrate type, chip size, ratio of analog-to-digital chip area, number of supply leads, and size of on-chip decoupling capacitors, as well as the type of noise that is the most problematic in each application [6.32, 6.33, 6.34, 6.35, 6.36, 6.37, 6.38, 6.39, 6.40, 6.41, 6.42, 6.43]. The adequate high-level substrate noise simulations are normally preferred for adjusting each of the above parameters to minimize performance degradation caused by substrate-coupled noise.

Since the used CMOS technology process is a lightly-doped substrate, unlike another common heavily-doped (low-ohmic) CMOS Epitaxial process, that can be simply modeled by a single-node bulk [6.32, 6.44], more sophisticated but complex modeling is normally needed by using specific CAD tools and extracted from layout for the 3-D admittance matrices [6.44, 6.45, 6.46, 6.47, 6.48, 6.49, 6.50, 6.51, 6.52]. Fortunately, the guard rings (provide noises a low resistance path to AC ground with thus minimized injection to the substrate) and "Moat" separation by NWELL (break current flow in the channel stop implant region) are more effective ways to reduce the substrate noise coupling in lightly-doped bulk process due to the fact that sufficient amount of the substrate current flows near the die surface because of the p+ channel stop implant [6.32, 6.33, 6.34, 6.35, 6.36, 6.39, 6.40, 6.44, 6.45, 6.49]. Therefore the circuit is carefully laid out with the following guidelines: the closed-loop guard-rings are placed around all analog cells with every tens of μm width rectangular region; NWELL barriers biased at VDD and with also closely surrounded p+ substrate tie biased to ground (achieve good supply decoupling at the same time) are also placed among all sensitive analog cells, like opamps, and between analog and digital region; ample substrate contacts with well ground biasing are placed in any unused space.

Hence, to reduce the digital supply noise, the dominant source for substrate noise coupling, is the most important key. Minimization of the current spikes flowing through inductive package is the main and effective way to reduce the dI/dt noise. On-chip decoupling capacitor is the simplest and very effective way for providing low-impedance path to current spike [6.33, 6.44, 6.53, 6.54]. The only caution is to avoid the LC resonant frequency located at signal band and main clock frequency. The increase of decoupling capacitance results in the reduction of the resonant frequency but more usefully also the Q factor, so the peaking and oscillation won't be very significant if having enough capacitance associated with the parasitic resistances of the power line. Moreover, the area price can be alleviated in modern technology containing 3 or more metal layers, as the decoupling capacitors normally formed by MOS transistors [6.33, 6.53, 6.54] can be placed under the power bus which is also required to be wide for high-frequency applications.

Nevertheless, the effectiveness of the on-chip decoupling capacitor approach indeed relies on the spike-current flow path. Although the multiple clock phases generate more digital noise than normal bi-phase SC circuit, the largest amount of current spikes is still originated from the clock buffering network driving the clock bus and switches. This implies that a correct and minimization of spike-current return paths from both digital (buffer transistor and clock bus parasitics) and from analog domain (switches) is the crucial and direct concern in coping with the on-chip decoupling.

The most common way of biasing the substrate for lightly-doped process are (1) substrate contacts on all power carrying grounds, i.e. separate analog and digital ground, or (2) substrate contacts on dedicated single Kelvin ground (non-power carrying) [6.32, 6.33, 6.34, 6.35, 6.36, 6.39, 6.45]. These can usually also be associated with dedicated power pins for the last digital buffer stage. However, from no matter the modeling and circuit simulation results, the above strategies lead to both larger on-chip supply noise when compared with that obtained using a simple single-shared-ground pin approach for both analog and digital, as well as substrate biasing. The main reason is due to the fact that the shared ground approach used here achieves perfectly the purpose of forcing the most spike-current flowing inside the chip, thus minimizing the dI/dt noise. Figure 6.19 presents clearly the return path of the main spike-current flow, imposed by the digital buffer driven for rising and falling edges, within both digital and analog supply domains when having a shared ground.

The shared-ground approach when compared with the separated ground scheme can reduce the supply noise by nearly 50 %, as confirmed by the simulations. Similar results have also been obtained in [6.40, 6.41, 6.42, 6.43]. Hence, MOS capacitors are filled in any unused space and also under the wide-sheet power supply lines for obtaining enough decoupling capacitance from the supplies to ground, and the estimated total capacitance value is more than hundreds of pF. Using small unit-capacitor array instead of large area MOS capacitor is preferred for better decoupling performance, by minimizing series impedance due to the parallelism nature. Furthermore, a large capacitor formed by nMOS capacitor array is also placed at VCM node to minimize the sampling glitches and inductive ringing, as well as to perfectly refer to ground node voltage, thus ensuring to the opamp a more stable negative swing margin which is the dominant factor for the opamp linearity due to the low common-mode level.

Note that this single-shared ground scheme implies also that all the signals inside the chip now refer to the same ground even if the ground is not clean when compared to the outside PCB ground for testing, as this injects only the common-mode noise to the signals that will be greatly attenuated by fully-differential scheme.

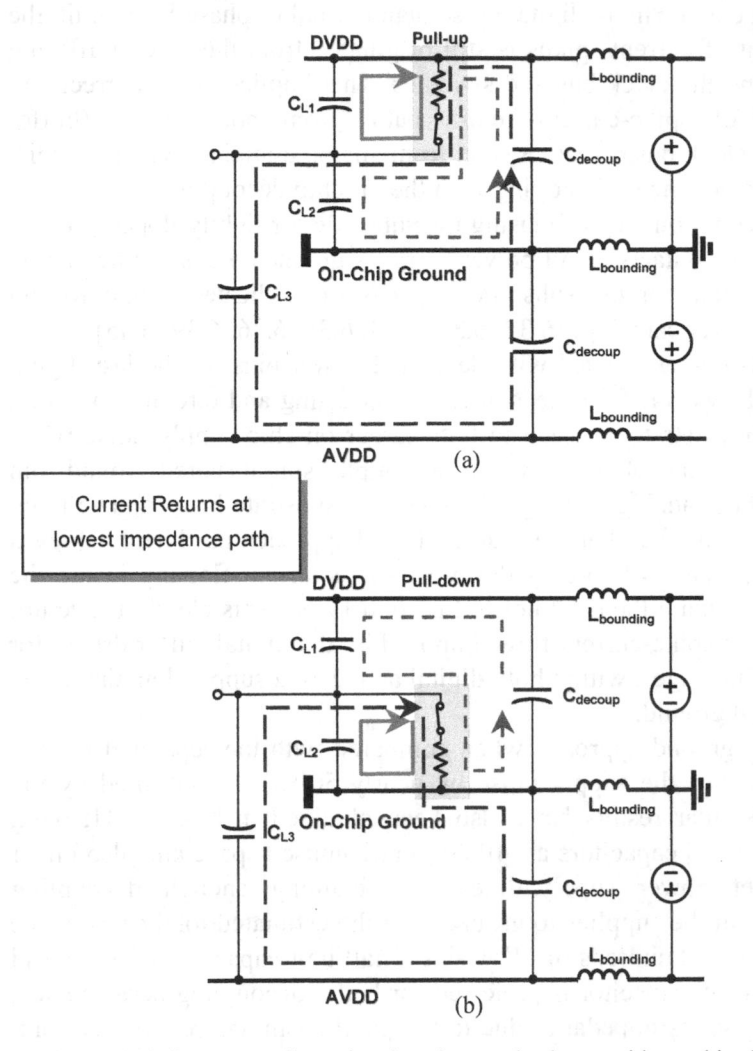

Figure 6-19. Spike-current flows for shared ground scheme with on-chip decoupling in (a) rising (b) falling edges

Moreover, shielding the long noisy clock phase distribution bus by digital VDD gets further 20% reduction in supply noise, in simulations, when compared to shielding to ground, since this avoids the maximum spike-current peak flowing into the common ground from the buffers through the clock bus parasitics. This is due to the fact that the maximum peak spike-current happens at the rising edges of clock phases, since all synchronized clock phases are raised simultaneously but do not fall down at the same time because of the delayed clock phase or bottom sampling techniques used for reducing charge-injection and clock-feedthrough effects.

4.3 Shielding

Shielding is also another very important issue for protecting the sensitive signal path, isolating the noisy signals, as well as improving the path matching. It is worth to point out here several details employed in the circuit layout:

- Multi-dimension shielding by lower, same and upper layers, are used if the area and resulting increased parasitic capacitance is allowed, e.g. opamp input nodes as shown in Figure 6-17, path isolation shielding as shown in Figure 6-16 and Figure 6-18. Either NWELL or poly layer are placed under signal lines and capacitor group.
- The shielding layers should be biased to the local VDD that supplies the device which generates the signal being shielded so as to inject as less as possible current into the common ground, i.e. shielding clock bus by metal 1 connected to digital VDD. However, if the shielding is used for isolating signals from two supply domains, the shielding layer should be biased to the common ground, e.g. for the case when analog signals cross over the digital clock signals.
- Use the most conductive layer available for shielding with also importantly ample contacts and vias to supply for reducing the RC time constant effect that will significantly limit the practical effect of shielding, e.g. use conductive metal 1 instead of NWELL or poly for shielding clock bus so that the noisy current spikes return immediately and straightforwardly with maximizing the cleanness of the environment; when using NWELL for shielding, the diffusion whose sheet resistance is more than 10 times less than that of NWELL in this process is also placed for reducing the series impedance inside the NWELL region.

4.4 Floor Plan

The floor plan and die microphotograph is presented in Figure 6-20. The input signals to the chip enter in the upper left corner and feed into the T/H stage clocked with the lower input sampling rate, and the T/H outputs are processed by the ADB time-interleaved serial delay line located on the top, the polyphase subfilters are placed in parallel in the left bottom. The clock generation is at the top-right corner, so that the noisy clock bus can be minimized and placed in middle. A perfect shielding isolation is used to deal with the unavoidable cross over between analog and digital signals, e.g. metal 1 for analog, metal 3 for digital and metal 2 for shielding with well biased to the common ground. As mentioned before, to minimize the mismatches among skew-sensitive time-interleaved phases, the high-speed clock generation part drives directly to the multiplexer, as located on the rightmost part of the chip. The total active area is about 2 mm² including the analog filter core, clock generator and input T/H stage [6.55, 6.56].

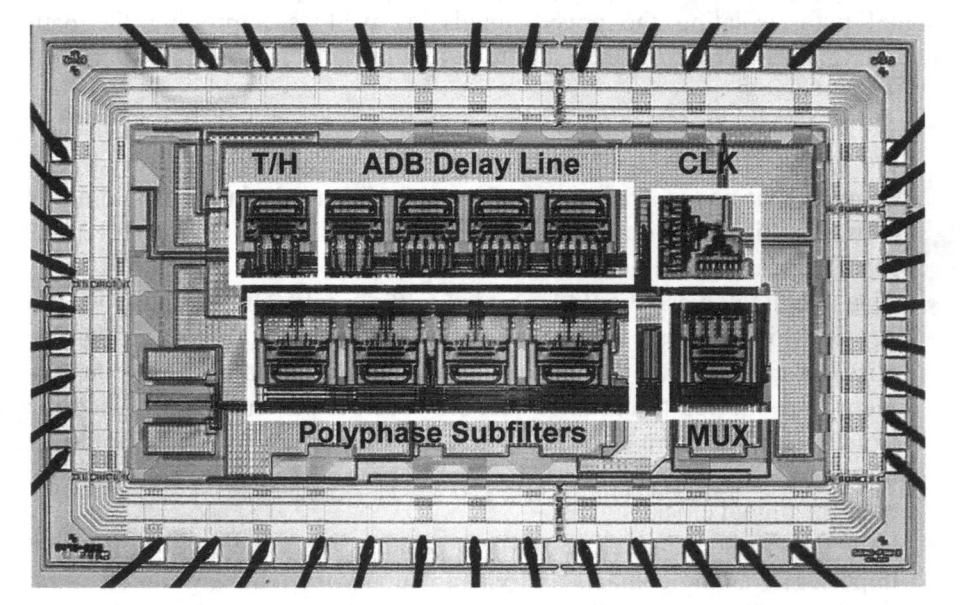

Figure 6-20. Die microphotograph

5. SIMULATION RESULTS

5.1 Opamp Simulations

The effectiveness of the designed Telescopic opamps, which is the heart of the SC filter, has been verified comprehensively by Cadence Spectre simulations by taking into account the physical process variations. Post-layout simulations show that the fastest opamp dissipates around 12mW at 2.5 V supply to achieve 66.6 dB gain, 1 GHz unit-gain bandwidth, with 63° phase margin for a 2.5 pF loading and about 1.6 ns to settle at 0.5 V voltage step with 1.7 V/ns slew rate. Figure 6-21 and Figure 6-22 present the AC open-loop and DC output swing of the parasitic-involved layout-extracted opamp from both the typical, and the worst-case corner process simulations. However, the worst-case corner is quite rare to happen, Monte-Carlo simulation is the best and indeed the most realistic way to foresee the opamp performance degradation to the process variations although it is computer time-consuming. Figure 6-23 shows the statistical Histogram of the 500-run simulation for unity-gain bandwidth, phase margin, DC gain as well as DC gain at 1.2 V_{p-p} differential swing, which all meet well the required specifications. Figure 6-24 also presents scatter plot from the same Monte-Carlo simulations giving a clearer picture of opamp performance. In addition to see how the switches' on-resistance in feedback path degrades the opamp close-loop stability, the Figure 6-25 shows the loop gain of the opamp and feedback path, illustrating clearly the reduction in the phase margin due to the presence of switches.

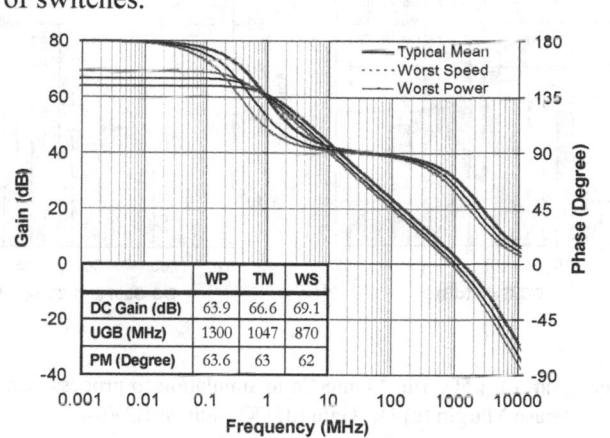

Figure 6-21. Opamp layout-extracted AC open-loop frequency response from corner simulations

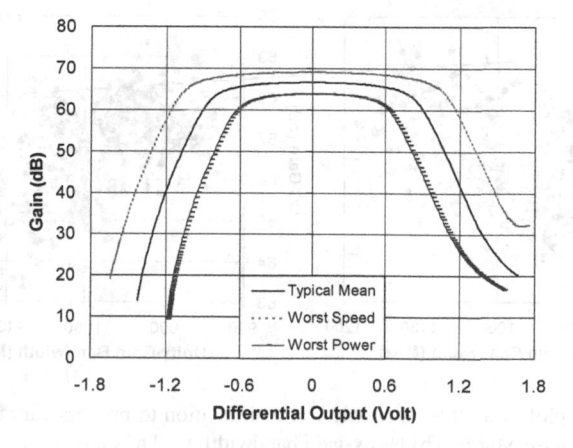

Figure 6-22. Opamp layout-extracted DC gain and output swing from corner simulations

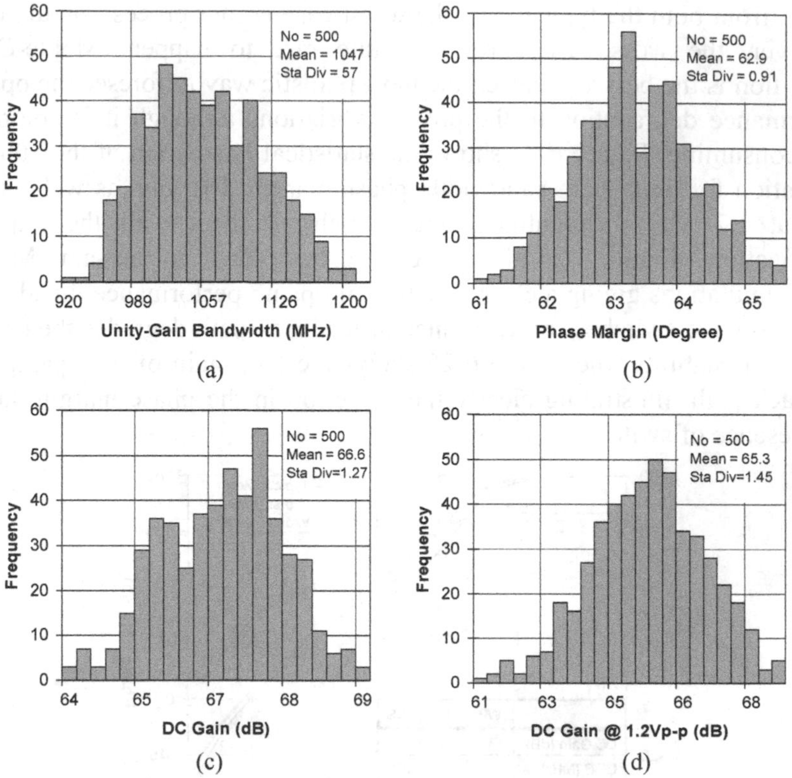

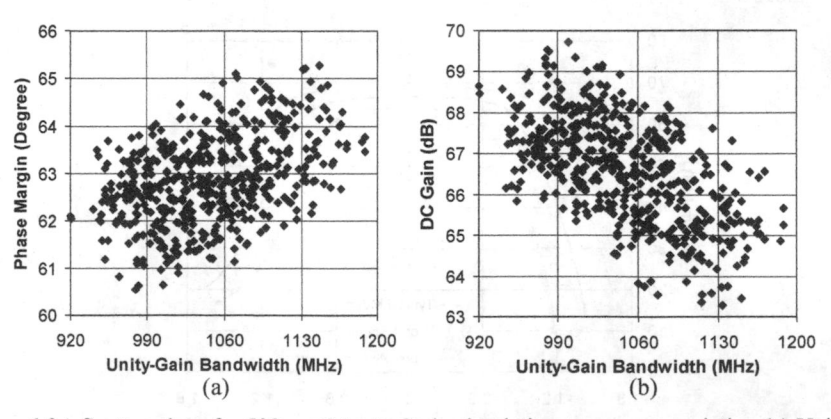

Figure 6-23. Histogram of a 500-run Monte-Carlo simulation to process variation (a) Unity-gain bandwidth (b) Phase Margin (c) DC Gain (d) DC Gain @ 1.2Vp-p

Figure 6-24. Scatter plot of a 500-run Monte-Carlo simulation to process variation (a) Unity-gain bandwidth vs. Phase Margin (b) Unity-gain bandwidth vs. DC Gain

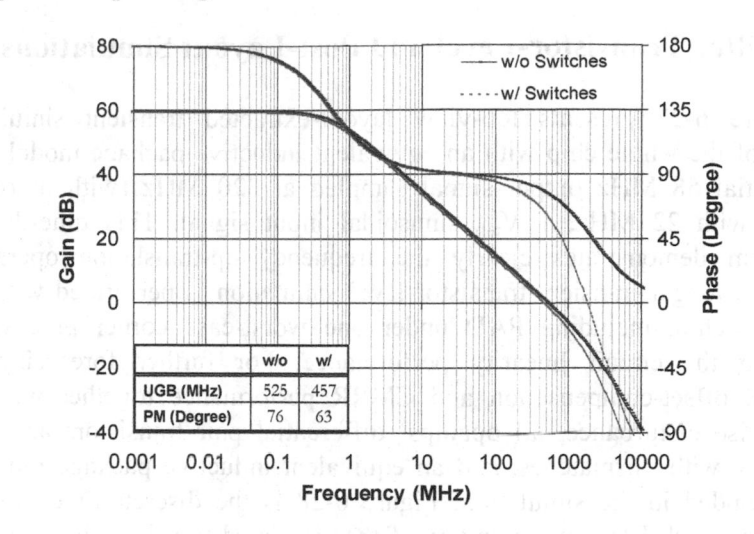

Figure 6-25. Opamp layout-extracted loop-gain with / without switch resistance in feedback path

5.2 Filter Behavioral Simulations

Figure 6-26 shows the Monte-Carlo amplitude response simulation with respect to all capacitance ratios, which are independent zero-mean Gaussian random variables with a large deviation within 2.1 % (or $\sigma = 0.7$ %), showing that the worst case of the images stopband is below −40 dB.

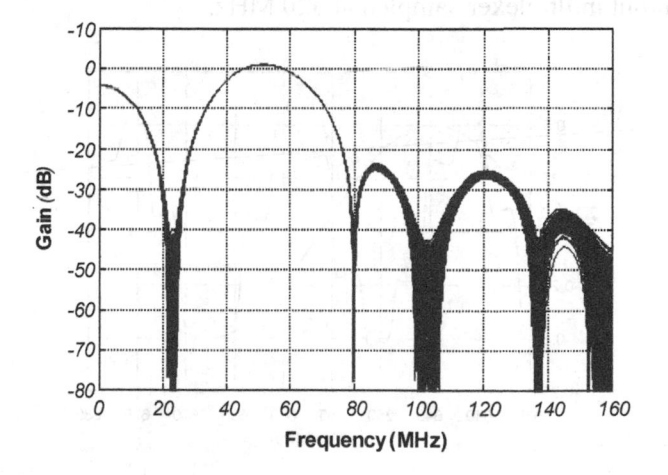

Figure 6-26. Monte-Carlo amplitude response simulations ($\sigma_e = 0.7$ %)

5.3   Filter Transistor-Level and Post-Layout Simulations

Figure 6-27 presents top-view layout-extracted transient simulation results of the whole chip with an equivalent inductive package model for a differential 58 MHz output signal sampled at 320 MHz (without output buffer) with 22 MHz, 1 V_{p-p} sinusoidal input signal. This time-domain waveform demonstrates clearly the frequency up-translation operation. Beside, a longer transient transistor-level simulation is performed with full top-view chip, including PAD under one worst-case corner process for verifying the circuit linearity performance. For further foreseeing the practical offset-compensation and CMRR performance together with the dI/dt noise disturbance, all opamps' differential pair transistors are sized randomly with mismatches, and an equivalent inductive package model is also included in the simulation. Figure 6-28 is the discrete-time Fourier-Transform with Blackman window of the simulated transient signal output, showing that the overall performance meets well with the design specifications: the input baseband signal at 22 MHz and its images at 102 MHz, 138 MHz have been attenuated all below 45 dB, the pattern-noise tone at 80 MHz and 160 MHz are mainly due to the assigned mismatches of opamp which is about -63 dBc. All the other tones are the frequency translated-images of the harmonics sampling at lower input and higher output sampling rate and their multiples, and the total harmonic distortion is calculated to be -56 dB. The worst harmonic tone is -57 dBc located at 146 MHz which is the addition of the 3rd-harmonic of signal 22 MHz from low-speed filter core sampled at 80 MHz and 3rd-hamonic of the output 58 MHz from the output multiplexer sampled at 320 MHz.

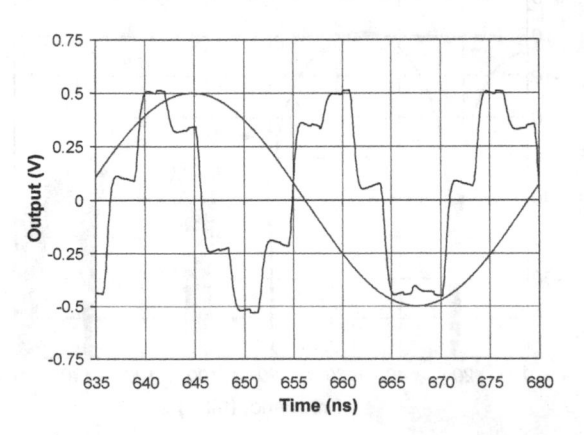

Figure 6-27. 58MHz output signal with a 1V_{p-p} 22MHz input (f_s=320MHz) from top-view layout-extracted simulation

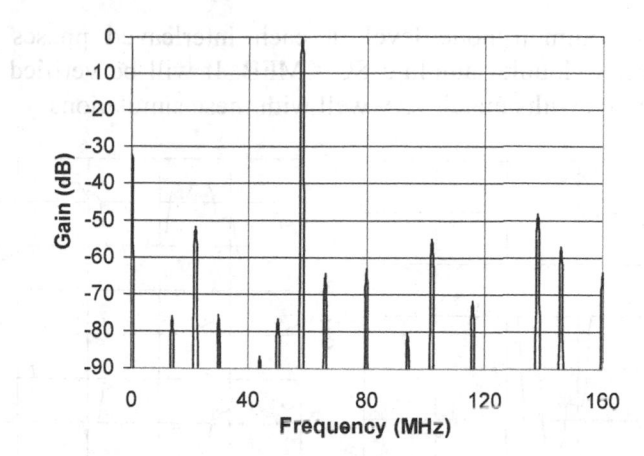

Figure 6-28. Spectrum of 58MHz output signal with a $1V_{p-p}$ 22MHz input (f_s=320MHz) from worst-case top-view transistor-level simulations

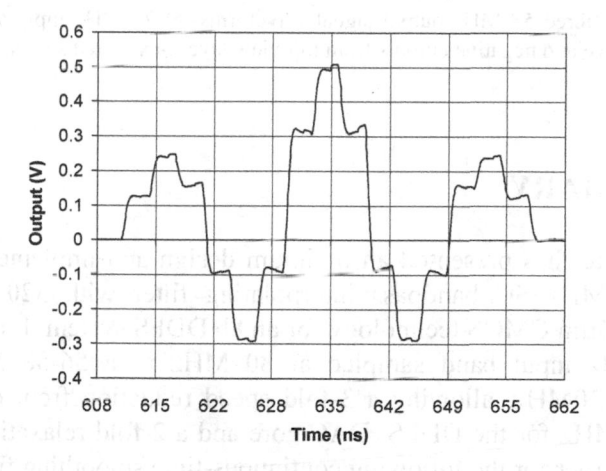

Figure 6-29. Impulse transient response from top-view layout-extracted worst-case simulation

Figure 6-29 presents the impulse transient response of the circuit obtained from the top-view layout-extracted simulation with also inductive package model under the worst-case speed and temperature. It shows clearly the achieved 15-tap symmetrical impulse response of the FIR filtering.

Finally, for comparison purpose with the measurements, Figure 6-30(a) presents the top-view layout-simulated 22 MHz input and 58 MHz output from the last level-shifter and buffer of the circuit (output from inductive package and PAD) that will be really measured by the equipment. Figure 6-30(b) is the detailed positive and negative output with also its CM signal which shows common-mode level in each interleaved phases correctly controlled by the double-sampling SC CMFB. It will be verified later that the experimental results match very well with these simulations.

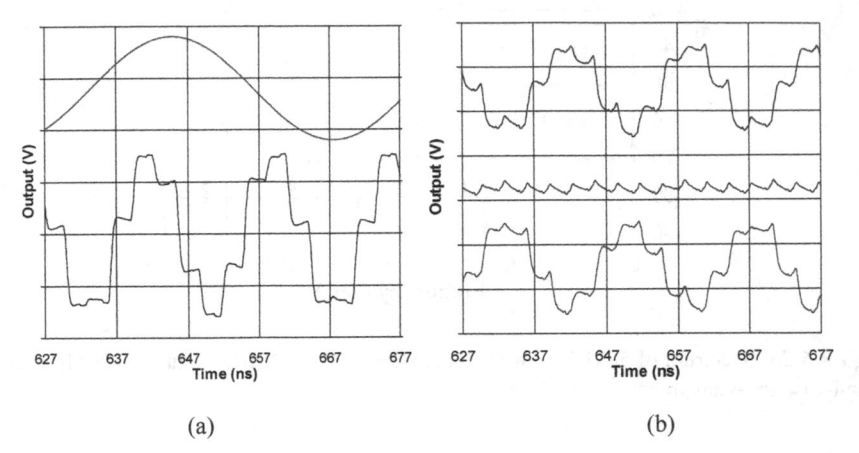

(a)   (b)

Figure 6-30. Buffered 58 MHz output signal waveforms (a) 22 MHz input and differential output (b) Positive and negative outputs from top-view layout-extracted simulations

6.   SUMMARY

This chapter has presented an optimum design and implementation of a 15-Tap, 57 MHz SC bandpass interpolating filter with 320 MSample/s output in 0.35μm CMOS technology for an 8b DDFS system. It up-translates a 22-24 MHz input band sampled at 80 MHz to a 56-58 MHz output sampled at 320MHz, allowing a 3-fold speed reduction from original 240 MHz to 80 MHz for the DDFS+DAC core and a 2-fold relaxation from 4th-order to 2nd-order for the following continuous-time smoothing filter.

Specific multi-notch FIR transfer function is designed for this narrow band filtering to avoid the use of a higher-sensitivity, high-Q IIR and rather large weight-spread, higher-order standard bandpass FIR function. A time-interleaved serial ADB delay line structure has also been customized to achieve the 15 tap delays with only 4 stages for minimum power and error accumulation. Autozeroing technique is employed in the low-speed filter core to alleviate the pattern noise imposed by the offset serial propagation and parallel mismatches. The mismatch-free property used in both the ADB and output high-speed multiplexer eliminates capacitance ratio mismatches, reduces finite gain errors, as well as boosts the achievable speed.

Practical prototype circuit implementation for each building block and the whole circuit has been comprehensively investigated including high-speed Telescopic opamp with wide-swing and internal biasing, SC CMFB, switches, capacitor sizing for each stage, noise contributions as well as I/O circuitry. To attain the stringent timing-skew requirement, a low phase-skew clock generation has been designed by focusing mainly on three controls: Design systematic mismatches control (Equal-propagation-gate-delay), Random process mismatches control (Output rising-edge synchronization) and dI/dt supply noise mismatch control (Separation of digital VDD supplies).

Special layout techniques for the whole chip have been studied to deal with the matching, parasitic and noise coupling problems. Common-centroid with dummy peripheral and mirror-symmetry are applied to the mismatch-sensitive part. Separated VDD supply pins but single-shared ground with on-chip decoupling for analog and digital parts have been proposed to reduce maximally the supply and substrate noise imposed by the inductive package when comparing to other grounding schemes. A clean signal and substrate environment is further achieved by ample substrate contacts and multi-dimensional shielding with minimized return-current-path impedance.

Behavioral, transistor-level and layout-extracted/parasitic-involved simulations have been performed for verifying the effectiveness of the proposed circuit techniques. The results show that the circuit functions well taking into account the worst-case process variations.

REFERENCES

[6.1] J.E.Franca, D.G.Haigh, "Design and Applications of Single-Path Frequency-Translated Switched-Capacitor Systems," *IEEE Trans. Circuits and Systems*, Vol.35, No.4, pp.394-408, Apr.1988.

[6.2] D.H.Shen, C-M.Hwang, B.B.Lusignan, B.A.Wooley, "A 900-MHz RF Front-End with Integrated Discrete-Time Filtering," *IEEE J. Solid-State Circuits*, Vol.31, No.12, pp.1945-1954, Dec.1996.

[6.3] P.J.Chang, A.Rofougaran, A.A.Abidi, "A CMOS channel-select filter for a direct-conversion wireless receiver," *IEEE J. Solid-State Circuits*, Vol.32, pp.722-729, May 1997.

[6.4] R.F.Neves, J.E.Franca, "A CMOS Switched-Capacitor Bandpass Filter with 100 Msample/s Input Sampling and Frequency Downconversion," in *Proc. European Solid-State Circuits Conference (ESSCIRC)*, pp.248-251, Sep.2000.

[6.5] Yi-Huei Chen, Jenn-Chyou Bor, Po-Chiun Huang, "A 2.5 V CMOS Switched-Capacitor channel-select filter with image rejection and automatic gain control," in *IEEE Radio Frequency Integrated Circuits (RFIC) Symposium Digest of Papers*, pp.111-114, 2001.

[6.6] L.K.Tan, H.Samueli, "A 200 MHz Quadrature digital Synthesizer/Mixer in 0.8μm CMOS," *IEEE J. Solid-State Circuits*, vol.30, No.3, pp.193-199, Mar.1995.

[6.7] A.Edman, A.Björklid, I.Söderquist, "A 0.8μm CMOS 350 MHz Quadrature Direct Digital Frequency Synthesizer with Integrated D/A Converters," *Proc. IEEE 1998 Symposium on VLSI Circuits Digest of Technical Papers*, pp.54-55, 1998.

[6.8] J.Vankka, M.Waltari, M.Kosunen, K.A.I.Halonen, "A Direct Digital Synthesizer with an On-Chip D/A-Converter," *IEEE J. Solid-State Circuits*, vol.33, No.2, pp.166-176, Feb.1998.

[6.9] A.Madisetti, A.Y.Kwentus, A.N.Willson, Jr. "A 100-MHz, 16-b, Direct Digital Frequency Synthesizer with a 100-dBc Spurious-Free Dynamic Range," *IEEE J. Solid-State Circuits*, vol.34, No.8, pp.1034-1042, Aug.1999.

[6.10] S.Mortezapour, E.K.F.Lee, "Design of Low-Power ROM-Less Direct Digital Frequency Synthesizer Using Nonlinear Digital-to-Analog Converter," *IEEE J. Solid-State Circuits*, vol.34, No.10, pp.1350-1359, Oct.1999.

[6.11] J.Jiang, E.K.F.Lee, "A ROM-less Direct Digital Frequency Synthesizer Using Segmented Nonlinear Digital-to-Analog Converter," *Proc. IEEE Custom Integrated Circuits Conference*, pp.165-168, May 2001.

[6.12] Y.P.Tsividis, "Integrated continuous-time filter design - An overview," *IEEE J. Solid-State Circuits*, vol.29, No.3, pp.166-176, Mar. 1994.

[6.13] N. Rao, V. Balan and R. Contreras, "A 3V 10-100-MHz Continuous-Time Seventh Order 0.05° Equiripple Linear Phase Filter", in *ISSCC Digest of Technical Papers*, pp. 44-46, Feb.1999.

[6.14] R.Castello, I.Bietti, F.Svelto, "High-frequency filters in deep-submicron CMOS technology," in *ISSCC Digest of Technical Papers*, pp74-75, Feb.1999.

[6.15] José Moreira, *Design Techniques for Low-Power, High Dynamic Range Continuous-Time Filters*, Ph.D. Dissertation, Instituto Superior Técnico, Portugal, 1999.

[6.16] G. Groenewold, "Low-power MOSFET-C 120 MHz Bessel allpass filter with extended tuning range," *IEE Proc. Circuits, Devices and Sys.*, vol.147, no.1, pp. 28–34, Feb.2000.

[6.17] Y.P.Tsividis, "Continuous-time filters in telecommunications chips," *IEEE Communications Magazine*, pp.132-137, Apr. 2001.

[6.18] G.Bollati, S.Marchese, M.Demicheli, R.Castello, "An Eighth-order CMOS low-Pass filter with 30–120 MHz tuning range and programmable boost," *IEEE J. Solid-State Circuits*, Vol.36, No.7, pp.1056-1066, Jul.2001.

[6.19] A.Nagari, G.Nicollini, "A 3 V 10 MHz pseudo-differential SC bandpass filter using gain enhancement replica amplifier," in *ISSCC Digest of Technical Papers*, pp.52-53, Feb.1997.

[6.20] K.V.Hartingsveldt, P.Quinn, A.V.Roermund, "A. 10.7 MHz CMOS SC Radio IF Filter with Variable Gain and a Q of 55", in *ISSCC Digest of Technical Papers*, pp152-153, Feb.2000.

[6.21] J.E.Franca, R.P.Martins, "IIR Switched-Capacitor decimator building blocks with optimum implementation," *IEEE Trans. Circuits and Systems*, Vol. CAS-37, No.1, pp.81-90, Jan. 1990.

[6.22] R.Naiknawave, T.S.Fiez, "Automated hierarchical CMOS analog circuit stack generation with intramodule connectivity and matching considerations," *IEEE J. of Solid-State Circuits*, Vol.34, No.3, pp.304-317, Mar.1999.

[6.23] Seng-Pan U, R.P.Martins, J.E.Franca, "Design and analysis of low timing-skew clock generation for time-interleaved sampled-data systems," in *Proc. The 2002 IEEE International Symposium on Circuits and Systems (ISCAS)*, USA, May 2002.

[6.24] J.M.Cohn, D.J.Garrod, R.A.Rutenbar, L.R.Carley, "KOAN/ANAGRAM II: New tools for device-level analog placement and routing," *IEEE J. of Solid-State Circuits*, Vol.26, No.3, pp.330-342, Mar.1991.

[6.25] E.Malavasi, A.Sangiovanni-Vincentelli, "Area routing for analog layout," *IEEE Trans. Computer-Aided Design of Integrated Circuits and Systems*, Vol.12, No.8, pp.1186-1197, Aug.1993.

[6.26] E.Malavasi, D.Pandini, "Optimum CMOS stack generation with analog constraints," *IEEE Trans. Computer-Aided Design of Integrated Circuits and Systems*, Vol.14, No.1, pp.107-122, Jan.1995.

[6.27] J.D.Bruce, H.W.Li, M.J.Dallabetta, R.J.Baker, "Analog layout using ALAS," *IEEE J. of Solid-State Circuits*, Vol.31, No.2, pp.271-274, Feb.1996.

[6.28] D.A.Johns, K.Martin, *Analog Integrated Circuit Design*, John Wiley & Sons, Inc., 1997.

[6.29] R.J.Baker, H.W.Li, D.E.Boyce, *CMOS Circuit Design, Layout, and Simulation*, IEEE Press, 1997.

[6.30] B.Razavi, *Design of Analog CMOS Integrated Circuits*, McGraw-Hill, Inc., 2001.

[6.31] M.J.McNutt, S.LeMarquis, J.L.Dunkley, "Systematic capacitance matching errors and correlative layout procedures," *IEEE J. of Solid-State Circuits*, Vol.29, No.5, pp.611-616, May.1994.

[6.32] T.J.Schmerbeck, "Noise coupling in mixed signal ASICs," *Low-Power HF Microelectronics: A Unified Approach*, Chapter 10, IEE Press, 1996.

[6.33] AMS Analog Group, *Crosstalk in Mixed-Signal Systems*, AMS, 1996.

[6.34] T.Blalack, *Switching Noise In Mixed-Signal Integrated Circuits*, Ph.D. Dissertation, Stanford University, USA, 1997.

[6.35] B.R.Stanisic, N.K.Verghese, R.A.Rutenbar, L.R.Carley, D.J.Allstot, "Addressing substrate coupling in mixed-mode IC's: simulation and power distribution synthesis," *IEEE J. of Solid-State Circuits*, Vol.29, No.3, pp.226-237, Mar.1994.

[6.36] X.Aragonès, A.Rubio, "Experimental comparison of substrate noise coupling using different wafer types," *IEEE J. of Solid-State Circuits*, Vol.34, No.10, pp.1405-1409, Oct.1999.

[6.37] Y.Zinzius, E.Lauwers, G.Gielen, W.Sansen, "Evaluation of the substrate noise effect on analog circuits in mixed-signal designs," in *Proc. South Southwest Symposium on Mixed-Signal Design (SSMSD)*, pp.131-134, 2000.

[6.38] M.v.Heijningen, J.Compiet, P.Wambacq, S.Donnay, M.G.E.Engels, I.Bolsens, "Analysis and experimental verification of digital substrate noise generation for Epi-type substrates," *IEEE J. of Solid-State Circuits*, Vol.35, No.7, pp.1002-1008, Jul.2000.

[6.39] M.Ingels. M.S.J.Steyaert, "Design strategies and decoupling techniques for reducing the effects of electrical interference in mixed-signal IC's" *IEEE J. of Solid-State Circuits*, Vol.32, No.7, pp.1136-1141, Jul.1997.

[6.40] B.Nauta, G.Hoogzaad, "How to deal with substrate noise in analog CMOS circuits," in *Proc. European Conference on Circuits, Theory and Design (ECCTD)*, pp.12-1/6, 1997.

[6.41] K.Falakshahi, *High-Speed High-Resolution D/A Conversion in CMOS*, Ph.D. Dissertation, Stanford University, USA, 1999.

[6.42] M.Felder, J.Ganger, "Analysis of ground-bounce induced substrate noise coupling in a low resistive bulk epitaxial process: design strategies to minimize noise effects on a mixed-signal chip," *IEEE Trans. Circuits and Systems – II: Analog and Digital Signal Processing*, Vol.46, No.11, pp.1427-1436, Nov. 1999.

[6.43] P.Larsson, "Measurements and analysis of PLL jitter caused by digital switching noise," in *Proc. European Solid-State Circuits Conference (ESSCIRC)*, Sep.2000.

[6.44] D.K.Su, M.J.Loinaz, S.Masui, B.A.Wooley, "Experimental results and modeling techniques for substrate noise in mixed-signal integrated circuits," *IEEE J. of Solid-State Circuits*, Vol.28, No.4, pp.420-429, Apr..1993.

[6.45] K.Joardar, "A simple approach to modeling cross-talk in integrated circuits," *IEEE J. of Solid-State Circuits*, Vol.29, No.10, pp.1212-1219, Oct.1994.

[6.46] F.J.R.Clement, E.Zysman, M.Kayal, M.Declercq, "LAYIN: Toward a global solution for parasitic coupling modeling and visualization," in *Proc. IEEE Custom Integrated Circuits Conference (CICC)*, pp.537-540, May 1994.

[6.47] N.K.Verghese, D.J.Allstot, M.A.Wolfe, "Fast parasitic extraction for substrate coupling in mixed-signal ICs," in *Proc. IEEE Custom Integrated Circuits Conference (CICC)*, pp.121 -124, 1995.

[6.48] K.J.Kerns, I.L.Wemple, A.T.Yang, "Efficient parasitic substrate modeling for monolithic mixed-A/D circuit design and verification," *Analog Integrated Circuits and Signal Processing*, 10, pp.7-21, 1996.

[6.49] R.Gharpurey, R.G.Meyer, "Modeling and analysis of substrate coupling in integrated circuits," *IEEE J. of Solid-State Circuits*, Vol.31, No.3, pp.344-353, Mar.1996.

[6.50] N.K.Verghese, D.J.Allstot, M.A.Wolfe, "Verification techniques for substrate coupling and their application to mixed-signal IC design," *IEEE J. of Solid-State Circuits*, Vol.31, No.3, pp.354-365, Mar.1996.

[6.51] A.Samavedam, A.Sadate, K.Mayaram, T.S.Fiez, "A scalable substrate noise coupling model for design of mixed-signal IC's," *IEEE J. of Solid-State Circuits*, Vol.35, No.6, pp.895-904, Jun.2000.

[6.52] N.P.Van der Meijs, A.J.Van Genderen, F.Beeftink, P.J.H.Elias, *SPACE User's Manual*, Department of Electrical Engineering – Delft University of Technology, The Netherland. URL: http://cas.et.tudelft.nl/~space/space.html.

[6.53] L.K.Wang, H.H.Chen, "On-chip decoupling capacitor design to reduce switching-noise-induced instability in CMOS/SOI VLSI," in *Proc. 1995 IEEE International SOI Conference*, pp.100-103, Oct.1995.

[6.54] P.Larsson, "Parasitic resistance in an MOS transistor used as on-chip decoupling capacitance," *IEEE J. of Solid-State Circuits*, Vol.32, No.4, pp.574-576, Apr.1997.

[6.55] Seng-Pan U, R.P.Martins, J.E.Franca, "A 2.5 V, 57 MHz, 15-Tap SC bandpass interpolating filter with 320 MHz output sampling rate in 0.35mm CMOS," in *ISSCC Digest of Technical Papers*, Vol.45, pp380-381, San Francisco, USA, Feb. 2002.

[6.56] Seng-Pan U, R.P.Martins and J.E.Franca, "A 2.5V 57MHz 15-Tap SC Bandpass Interpolating Filter with 320MHz Output Sampling Rate in 0.35mm CMOS," *IEEE J. of Solid-State Circuits*, pp. 87-99, vol.39, January, 2004.

Chapter 7

EXPERIMENTAL RESULTS

1.   INTRODUCTION

The experimental prototype of the 15-tap 4-fold SC bandpass interpolating filter proposed in Chapter 6 has been fabricated in 0.35 μm double-poly, triple-metal CMOS process (AMS – Austria Mikro Systeme International AG). Totally 10 test chips were obtained with packaging in the 44-pin Ceramic Quad Flat-Pack (CQFP). Testing a SC filter with over hundreds of MHz output sampling rate entails many challenges, requiring great care in the design of the testing board and the measurement setup. This chapter will first present the design of a Printed-Circuit Board (PCB) with good Electromagnetic Compatibility (EMC) and the setup of the testing and characterization. The measurement results will then be illustrated, comprehensively, in order to consolidate the theoretical expectation, and finally a comparison will be drawn for the performance of the implemented filter versus the performance of the previously published high-frequency SC filters.

2.   PCB DESIGN

For operating frequency in the MHz range, it is not possible to use a length of track or wire (or star grounding) as a common ground. Only a solid area of conductor, namely ground planes, can provide a good ground up to 1 GHz (and beyond) with a dramatic reduction of all unwanted Electromagnetic (EM) interference [7.1, 7.2, 7.3, 7.4, 7.5]. Consequently, for correct characterization of this high-frequency bandpass SC interpolating filter operating at hundreds of MHz, a 4-layer PCB was carefully designed and implemented. The simplified functional block diagram of this PCB is shown in Figure 7-1, and the top view and bottom view of the PCB are presented in Figure 7-2. Design considerations and techniques for different blocks to achieve a better EMC will be presented next.

2.1   Floor Plan

The detailed description of each of these 4 layers in the PCB is shown in the Table 7.1. The signals are drawn on the top and bottom layer, the 2nd layer is the common ground plane and the 3rd layer is not only for the VDD line routing, but more importantly, most of the region of this layer is also used as ground, especially the area under the Device Under Test (DUT) coving the decoupling networks that concentrates the majority of the current flow. For having further better grounding, that is the main part of this application, all the empty space in top and bottom layer is filled with ground metal, with enough vias for good contacts to the 2nd common ground layer.

Table 7-1. Signals in different layer of PCB

Top Layer	Signals
Internal Layer 1	Ground
Internal Layer 2	VDDs, Ground
Bottom Layer	Signals

10 chip prototypes were received and tested initially in a special Yamaichi QFP production-used socket [7.6] instead of the traditional Test-and-Burn-in socket, which normally has considerably large inductive leads that will dramatically degrade the circuit performance in this frequency. Moreover, one additional important advantage of this production-used socket is its identical footprint to the CQFP package, so that the final characterization of the chip can be done by soldering one prototype onside the PCB for an even more minimized leads' inductivity.

All the other devices, most of them being Surface-Mount Device (SMD) for better high-frequency performance, are efficiently located surrounding the socket which is placed as a star center of the PCB, as shown in Figure 7-2, to minimize the tracks length. The closest surroundings of the socket/chip are occupied by the power supply decoupling networks, analog input and output are drawn in the left side and the digital master clock input port is located on the opposite side of the analog signals. The regulated voltage and current reference generation are placed at the outmost part of the PCB.

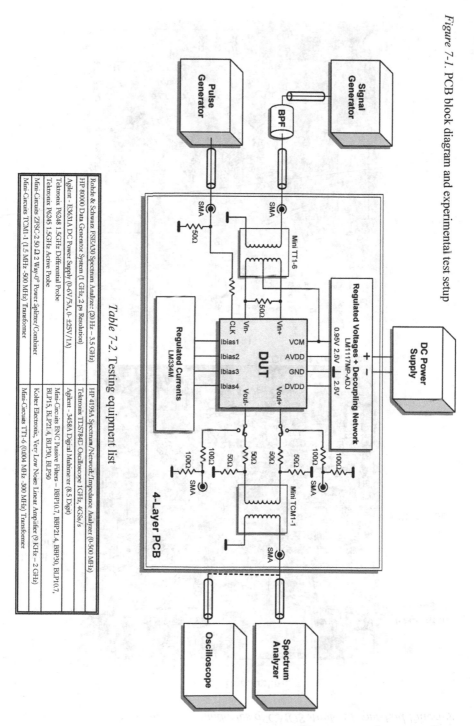

Figure 7-1. PCB block diagram and experimental test setup

Table 7-2. Testing equipment list

Rohde & Schwarz FSEA30 Spectrum Analyzer (20 Hz – 3.5 GHz)	HP 4195A Spectrum/Network/Impedance Analyzer (0-500 MHz)
HP 80000 Data Generator System (1 GHz, 2 ps Resolution)	Tektronix TDS784D Oscilloscope 1GHz, 4GS/s
Agilent – E3631A DC Power Supply (0-6V/5A, 0-±25V/1A)	Agilent – 3458A Digital Multimeter (8.5 Digit)
Tektronix P6248 1.5GHz Differential Probe	Mini-Circuits ENC Passive Filters – BBP10.7, BBP21.4, BBP30, BLP10.7, BLP15, BLP21.4, BLP30, BLP50
Tektronix P6245 1.5GHz Active Probe	Kolter Electronic, Very Low Noise Linear Amplifier (9 KHz – 2 GHz)
Mini-Circuits ZFSC-2-50 Ω 2-Way-0° Power Splitter/Combiner	
Mini-Circuits TCM1-1 (1.5 MHz -500 MHz) Transformer	Mini-Circuits TT1-6 (0.004 MHz -300 MHz) Transformer

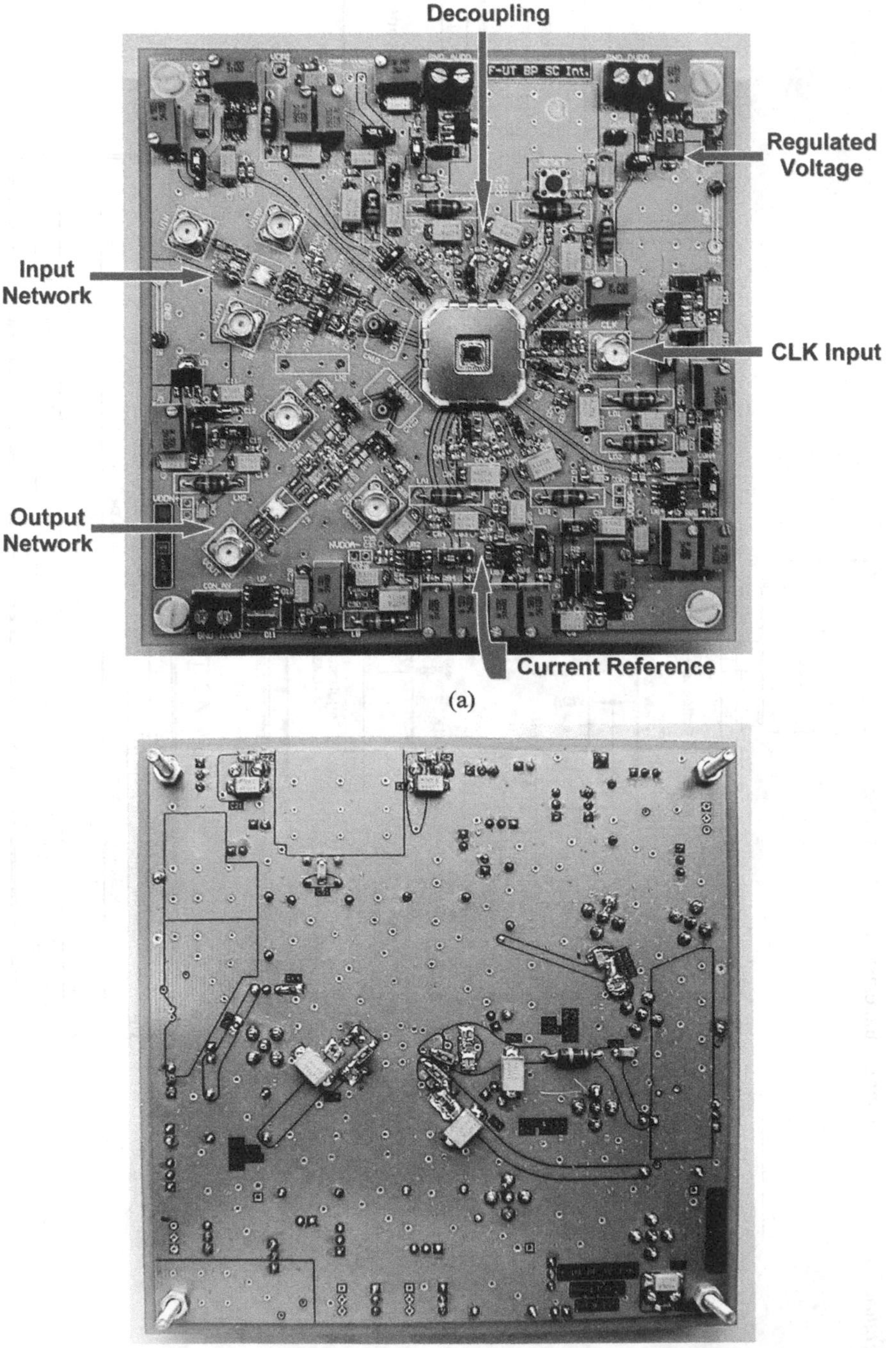

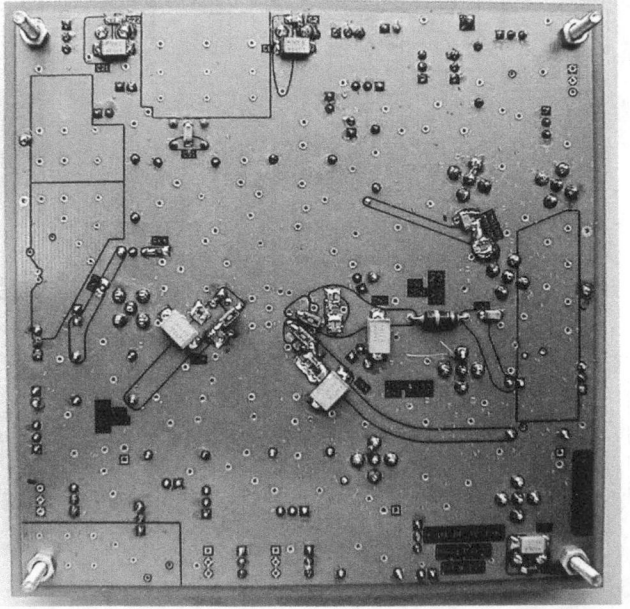

Figure 7-2. (a) Top-view (b) Bottom-view of the 4-layer PCB

2.2 Power Supplies and Decoupling

The supply voltages for both analog and digital parts of the chip are generated on the PCB to increase the cleanness of the environment (for testing purpose, these supplies can also be switched to external sources). The SMD low-dropout linear voltage regulator LM1117 is employed to generate 1.5-3.6 V adjustable voltage supply [7.7]. The standalone regulator provides different analog and digital supplies for the chip to minimize the noise coupling (however, no ill effect was observed in the measurement if only one regulator is used for all the supplies to the chip due to the good PCB board decoupling and also the on-chip decoupling).

The LC passive filtering forms the decoupling network where besides the tantalum μF SMD capacitor, there are 4 parallel multilayer ceramic SMD capacitors valued 100 nF, 10 nF, 1 nF and 270 pF, placed closely to the supply line at the nearest possible to the package supply pins, with also the dedicated vias linked to common ground. This combined parallel decoupling is employed so as to treat the self-resonant limitations of the capacitors, which leads to ineffectiveness of the low impedance path to the ground for the high-frequency noise [7.1, 7.2, 7.3, 7.4, 7.5, 7.8, 7.9, 7.10, 7.11]. To alleviate the parallel resonance problems possibly associated with this combining decoupling [7.2] and maintain a low level impedance across a very wide frequency range, these parallel-capacitor set are added at each power supply pin of the chip.

2.3 Biasing Currents

For testing purpose, four adjustable current references are needed to feed into the on-chip cascode current-mirror for distribution to the different blocks of the filter and buffer stages in order to have proper voltage biasing. They are generated by 3-terminal current source SMD LM334 [7.12] with the designed range of several tens of μA up to hundreds of μA. The current output lines are also decoupled to ground, for high-frequency noise elimination by parallel combination of the 10 μF, 0.1 μF and 1 nF SMD capacitors.

2.4 Input and Output Network

Since the filter is fully-differential, the signal conversions from single-ended to fully-differential and vice-versa are mandatory for I/O communication with the measurement equipment. To avoid any other noise source, the transformer-coupled I/O interfaces have been designed for signal conversion as shown in Figure 7-1 by using Mini-Circuit RF wideband transformers TT1-6 (0.004 MHz – 300 MHz) for input and TCM1-1 (1.5 MHz – 500 MHz) for output [7.13]. The common-mode level of the input signals is properly setup by connecting the center-tap of secondary side of the input transformer to the Voltage Common-Mode (VCM). For proper matching to the 50 Ω cabling and equipment, a 50 Ω resistor lies between transformer two-terminal outputs is used as input termination. Likewise, to drive the 50 Ω coaxial cable and input impedance of equipment for high-bandwidth measurements, the termination is achieved via a 50 Ω series back termination resistor. Note that the output voltage level will be halved in this scheme. Both the positive and negative terminal outputs of the filter can also be measured through the similar series back impedance termination. A 50 Ω resistor terminates the Clock input, and a series resistor in the clock input path damps the ringing caused by bounding wire inductance. SMA connectors are used for input and output.

Transmission lines are normally needed for signal integrity when PCB track is so long that a signal traveling along it cannot maintain the same potential at all points along its path due to its finite velocity of propagation. A rule-of-thumb is to treat PCB tracks using transmission line techniques, when connection length is greater than $\lambda/7$ to $\lambda/10$ at maximum of frequency interest [7.1, 7.2, 7.3]. Considering that the output signals are sampled and held at 320 MHz or higher, the imaging bands will be located at this sampling rate and its multiples, and the digital clock signal is also at this rate or higher, thus all the signal tracks are treated as transmission line. Transmission line in PCB normally can be described as following type: microstrip lines, coplanar waveguides, coplanar strips and conductor (ground)-backed coplanar waveguides, and their characteristic impedance can be controlled by the materials and dimensions [7.14, 7.15, 7.16, 7.17]. As the tracks are shielded in both sides and the lower ground plane, the determination of the characteristic impedance follows the formulas of conductor-backed coplanar waveguides [7.15, 7.16, 7.17]. According to the foundry-given relative permeability and thickness of the substrate, the characteristic impedance versus track width and gap/spacing between two-side ground planes is presented in the Figure 7-3. By choosing the width as 23 mil the gap is used at 11 mil in the entire signal track layout.

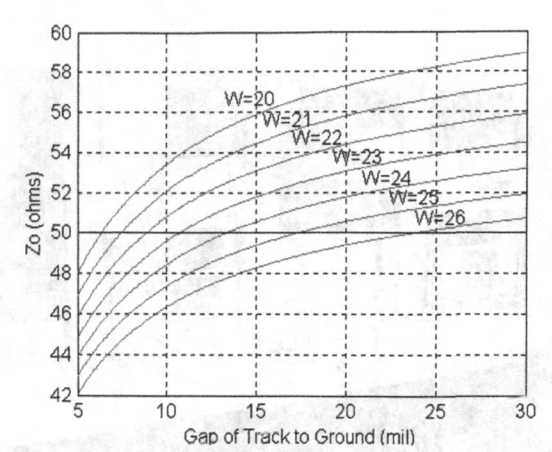

Figure 7-3. Characteristic impedance for conductor-backed coplanar waveguides versus track width and gap

3. MEASUREMENT SETUP AND RESULTS

The measurements are taken for 10 samples (only 10 from the foundry) at room temperature for different sampling rates and supply voltage. For simplicity, we present here the results for only the cases with output sampling rate at 160 MHz and 320 MHz with 2.5 V analog and digital supplies. Further results presented here are measured at 400 MHz output rate at 2.5 V analog and 3.3 V digital supplies. The characterization of chip mainly focuses on the frequency-domain measurement, e.g. amplitude response, group delay, Total Harmonic Distortion (THD), 3rd-order Intermodulation Distortion (IM3), noise, Common-Mode Rejection Ratio (CMRR), Power supply Rejection Ratio (PSRR) and so on. The time-domain test is also performed mainly for functional verifications.

Figure 7-1 shows the time-domain or one-tone spectrum measurement setup: input synthesizer sinusoidal output is first connected to a Mini-Circuit passive bandpass filter [7.13] to suppress the harmonics less than -70 dBc, this signal is then AC coupled by the transformer to convert from single-ended to fully-differential with common level at VCM. The output signal of the chip can be measured directly by the spectrum analyzer or observed in the oscilloscope by either passing by the 50 Ω matching network or using even wideband high-impedance differential probe. The measurement equipment list is presented in Table 7.2 and the view of the laboratory testing equipment is presented in Figure 7-4. The detailed measurement results are presented next.

Figure 7-4. View of laboratory testing instruments (Intermodulation distortion measurement)

3.1 Frequency Response

The measured amplitude response in Figure 7-5 shows that the minimum stopband rejections all satisfy the minimum specification of 40 dB stopband attenuation, i.e. >50 dB, 45 dB and 40 dB for 160 MHz, 320 MHz and 400 MHz output rates. Figure 7-6(a) presents the measured amplitude response of 10 samples at 320 MHz sampling rate simultaneously with the ideal response and simulated worst-case response, showing that the results meet well with the specifications. The passband ripple is smaller than 0.6 dB with 3σ at 0.02 dB for 10 samples due to the low passband sensitivity nature of FIR filtering. The variation in stopband attenuation of the filter has 3σ at only 0.45 dB in the 22-24 MHz band. In addition, the 10-samples amplitude response for 160 MHz and 400 MHz are also presented in Figure 7-6(b) and (c), respectively, which shows also relatively small variation to the process. Furthermore, due to the linear phase nature of this interpolating filter, the measured group delay variations within the interest passband is rather small, they are 6ps, 8ps, and 62 ps for 160 MHz, 320MHz and 400 MHz output rates, respectively.

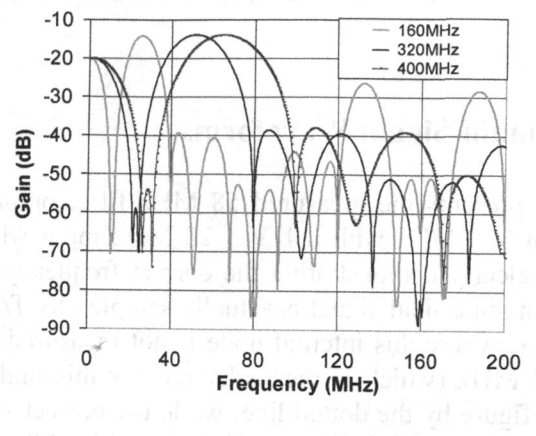

Figure 7-5. Measured amplitude responses for different output sampling rates

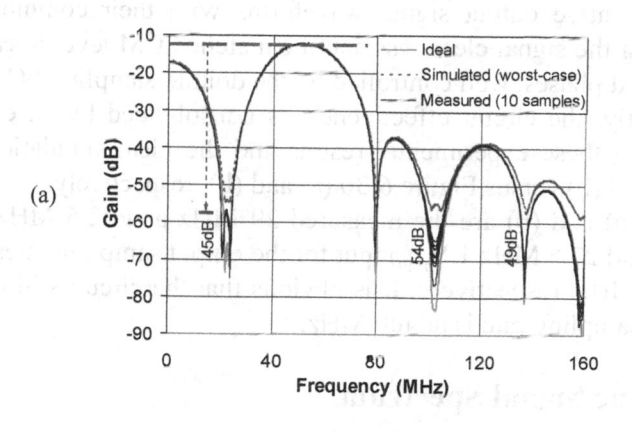

Figure 7-6. Measured amplitude response for 10 samples with (a) 320 MHz (b) 160 MHz (c) 400 MHz output sampling rates

3.2 Time-Domain Signal Waveforms

Figure 7-7(a) presents the measured 58 MHz filter output interpolated signal sampled at 320 MHz with a 1 V_{p-p} 22 MHz input where each time division is 5 ns, clearly demonstrating the correct frequency up-translation process. The input sinusoidal signal is actually sampled by T/H stage before inputting the filter, where this internal node is not measured, thus the ideal T/H signal at 80 MHz (which is the real input for this multirate filter) is illustrated in the figure by the dotted line, while the correct 4-fold sampling rate increase process is also illustrated clearly. In addition, to see the common-mode disturbed signals, Figure 7-7(b) presents the measured positive and negative output signal waveforms with their common-mode signals, showing the signal cleanness and the matched CM level in each one of the interleaved phases, well controlled by the double-sampling SC CMFB. More importantly, the circuit effectiveness is consolidated by an excellent matching among these experimental results and the ideal simulation ones: Figure 7-7(a) and (b) versus Figure 6-30 (a) and (b), respectively.

Figure 7-8 (a) and (b) are the measured 29 MHz and 72.5 MHz output with 11 MHz and 27.5 MHz 1 V_{p-p} input for the output sampling rates of 160 MHz and 400 MHz, respectively. It is obvious that the circuit still operates well when the sampling rate is at 400 MHz.

3.3 One-Tone Signal Spectrum

Figure 7-9(a) presents the measured spectrum of 58 MHz interpolated output sampled at 320 MHz with a 1 V_{p-p} 22 MHz input. The input signal and its sampling images at 80 MHz and its multiples, e.g. 102 MHz and 138 MHz, are attenuated correctly by a value greater than 45 dB, and the delta marker 1, 2 and 3 in Figure 7-9(a) are the folded images of their 3rd-harmonics sampled either at 80 MHz or 320 MHz, which are all less than -66.8 dBc, and the observed 2nd-harmonic is only as low as -78 dBc due to the careful layout. This results in a -62.4dB THD measured with 20 tones by counting up to 5th harmonic of both input and output signals as well as their folded images by the multirate sampling within the Nyquist band. Figure 7-9(b) presents the measured 1 % THD (-40dB) for a 2.1 V_{p-p} 22 MHz input, where the 3rd-harmonics are the dominant non-linearity distortion sources, and the 2nd-harmonics are still below -75 dBc.

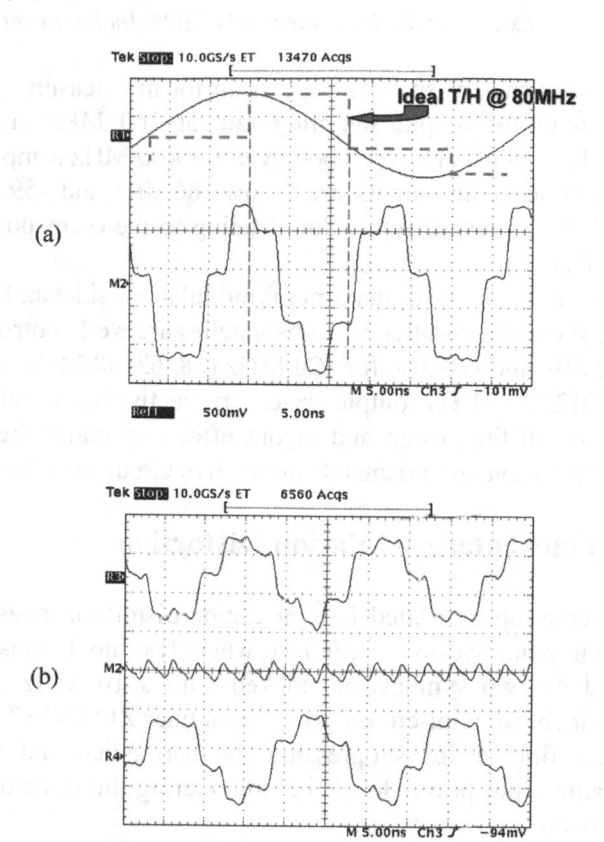

Figure 7-7. Measured 58 MHz output signal waveforms sampled at 320 MHz (a) 22 MHz input and differential output (b) Positive and negative outputs

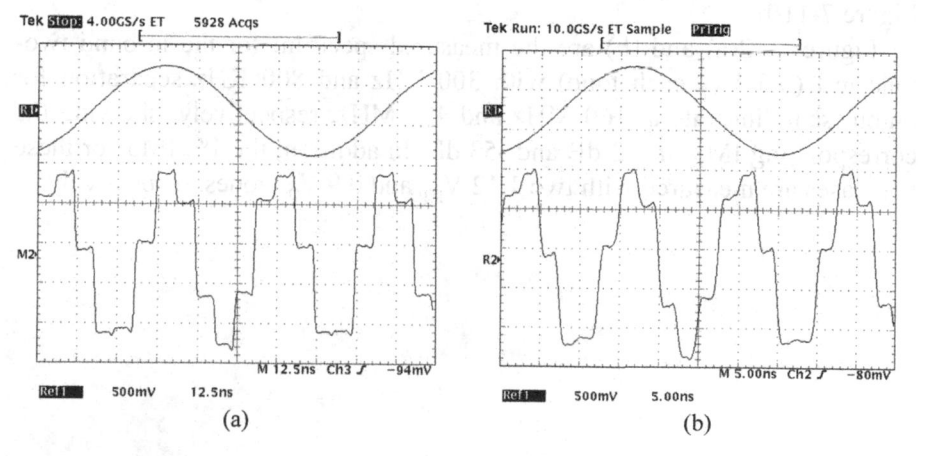

Figure 7-8. Measured signal waveforms (a) 11 MHz input, 29 MHz output for 160 MHz sampling rate (b) 27.5 MHz input, 72.5 MHz output for 400 MHz sampling rate

Figure 7-10(a) and (b) show the spectrum for the measured 29 MHz and 72.5 MHz output for output sampling rate at 160 MHz and 400 MHz, respectively. The results are very good even for 400 MHz sampling rate, e.g. all the 3rd-harmonic components are below -66 dBc and -59 dBc for 160 MHz and 400 MHz, respectively, thus leading to the corresponding -64 dB and -57 dB THD.

Especially, the observed maximum modulation sidebands due to the phase timing-skew and parallel gain mismatches are well controlled and only as low as -72 dBc and -70 dBc for 320 MHz (18, 62, 98 MHz) and 400 MHz (22.5, 77.5, 122.5 MHz) output rates, respectively, which shows the effectiveness of all the design and layout efforts specially the low timing-skew clock generation and mismatch-insensitive circuit structures.

3.4 Two-Tone Intermodulation Distortion

The frequency up-translated intermediation distortion measurements are fulfilled by a in-band two-tone input test, where two input sinusoidals, which are generated by two synthesizers locked with a 10 MHz reference, are added by a wideband Mini-circuit power combiner ZFSC-2 [7.13]. They are also bandpass filtered for suppressing the harmonics and measured (to ensure the same input power level) before entering the transformer-coupled signal conversion.

Figure 7-11(a) is the measured in-band spectrum of 56.7 MHz and 57.3 MHz outputs with two 0.5 V_{p-p} 22.7 MHz and 23.3 MHz inputs with the 320 MHz output sampling rate. The 3rd-order intermodulation distortion (IM3) is -52 dB, and 1% IM3 corresponds to two $0.85V_{p-p}$ tones, as shown in the Figure 7-11(b).

Figure 7-12(a) and (b) are the measured spectrum for the in-band two-tone test (0.5 V_{p-p} each tone) with 300 KHz and 800 KHz separation for output sampling rate at 160 MHz and 400 MHz, respectively, showing the corresponding IM3 of -62 dB and -53 dB. In addition, the 1% IM3 for these two cases are measured with two 1.12 V_{p-p} and 0.9 V_{p-p} tones, respectively.

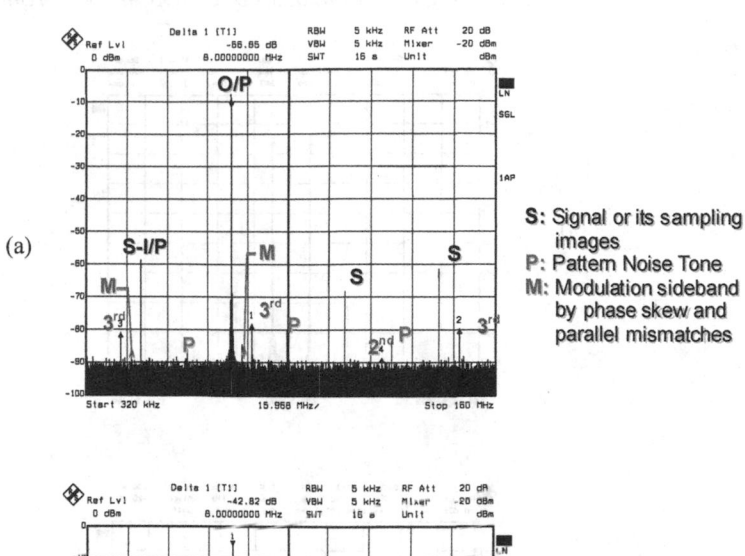

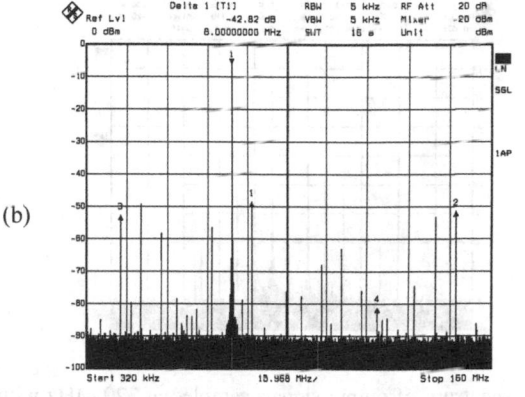

Figure 7-9. Measured spectrum of 58 MHz output signal sampled at 320 MHz with (a) 1 $V_{p\text{-}p}$ and (b) 2.1 $V_{p\text{-}p}$ 22 MHz input

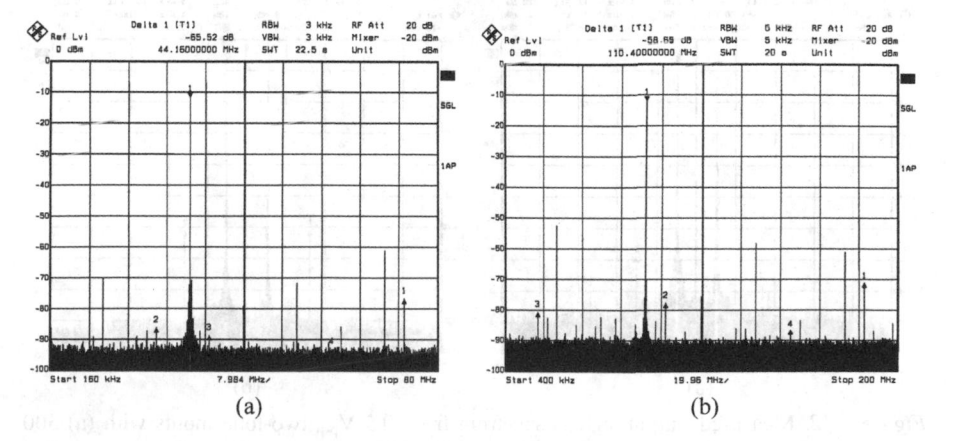

Figure 7-10. Measured signal spectrum (a) 29 MHz output for 160 MHz sampling rate (b) 72.5 MHz output for 400 MHz sampling rate

(a)

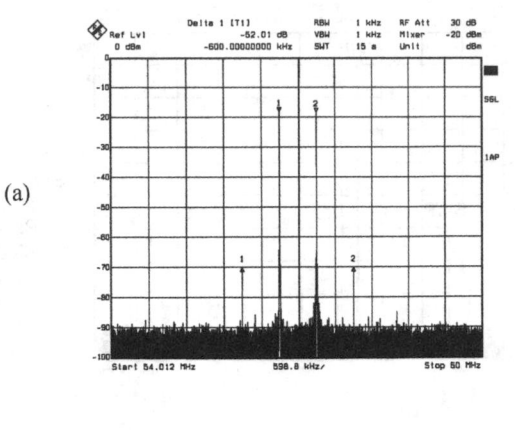

(b)

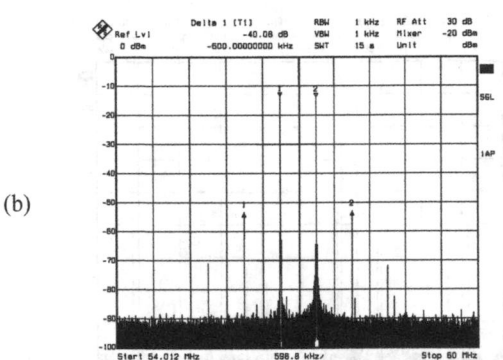

Figure 7-11. Measured spectrum of output signals sampled at 320 MHz with (a) 0.5 V_{p-p} and (b) 0.85 V_{p-p} two-tone inputs with 600 KHz separation

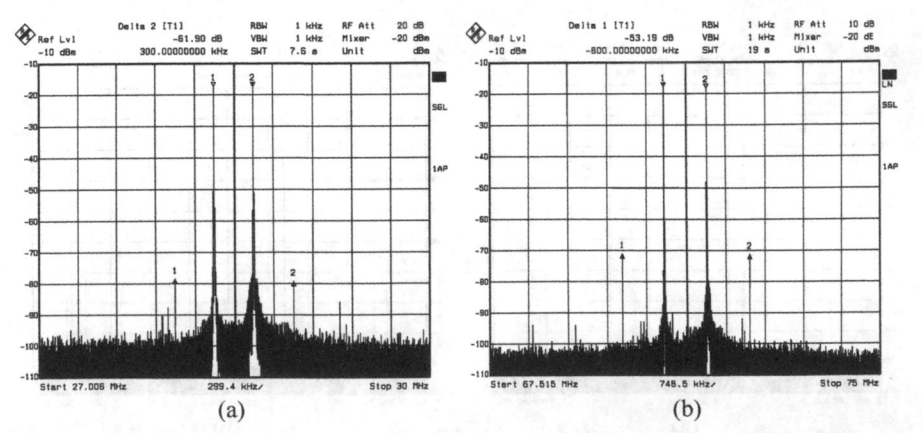

(a)   (b)

Figure 7-12. Measured output signals spectrum from 0.5 V_{p-p} two-tone inputs with (a) 300 KHz separation for 160 MHz sampling rate (b) 800 KHz separation for 400 MHz sampling rate

3.5 THD and IM3 vs. Input Signal Level

For evaluating circuit dynamic performance with respect to distinct input signal levels for different sampling rates, the above one-tone and two-tone tests are both performed to each case, and the Figure 7-13 reports the measured IM3 and THD performance versus different input signal levels for 160 MHz, 320 MHz and 400 MHz output rates. From the IM3 plot, the 3rd-order harmonics and input signal level follow well the theoretical expectation that 1 dB increase of input signal level will result in a 3 dB increase of the 3rd-harmonics, and the estimated output 3rd-order intercept points for 160 MHz, 320 MHz and 400 MHz sampling rates can be extrapolated from those measured data as 26.4 dBm, 23 dBm and 22.5 dBm, respectively.

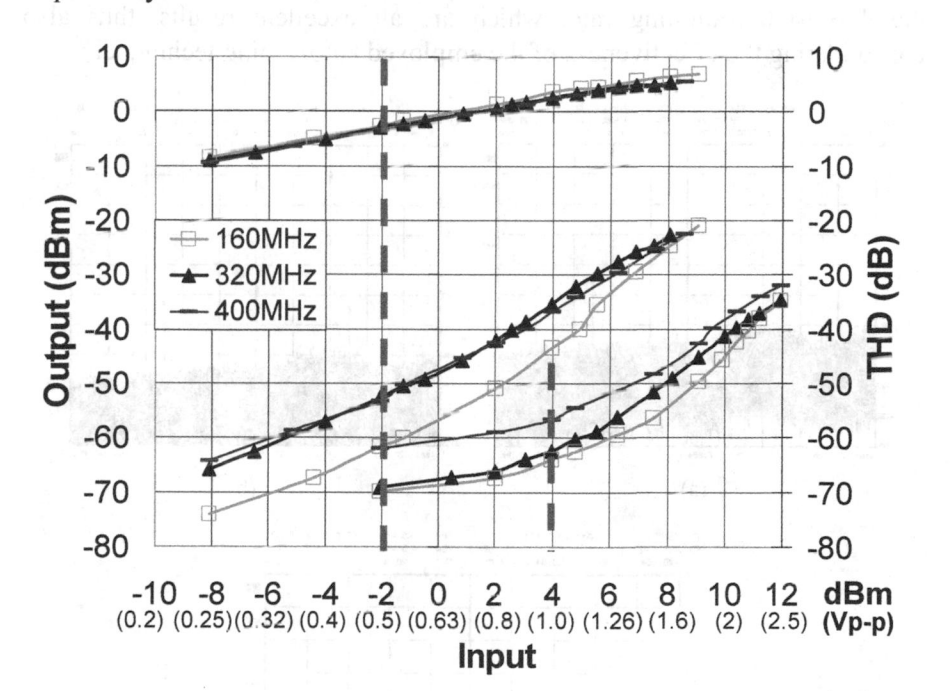

Figure 7-13. Measured THD and IM3 vs. input signal level for different output sampling rates

3.6 Noise Performance

Noise measurement contains two parts, fixed-pattern noise and noise floor. Figure 7-14(a), (b) and (c) are the fixed-pattern noise tones measured with zero input for 160 MHz, 320 MHz and 400 MHz sampling rates. The measured total fixed-pattern noise for circuit operating at 160 MHz and 320 MHz sampling rate are only as small as 120 μV_{rms} and it is doubled to 243 μV_{rms} for 400 MHz. This result verifies that fixed-pattern noise is mainly contributed from offset mismatches happened in the parallel signal paths due to the clock-feedthrough and charge-injection errors from the switches rather than the DC offset of opamp, because the digital supplies for the former two cases are the same at 2.5 V but it is 3.3 V for the last one (all the analog supply are at 2.5 V), and the increased clock phase voltage will certainly increase the clock-feedthrough and charge-injection errors from the switches due to the enlarged gate source/drain voltage. From Figure 7-9 and Figure 7-10, the maximum level of the pattern noise is 70 dB below the signal for the circuit with 160 MHz and 320 MHz sampling rate and it is still -65 dBc for the 400 MHz sampling rate, which are all excellent results, thus also consolidating the effectiveness of the employed autozeroing technique.

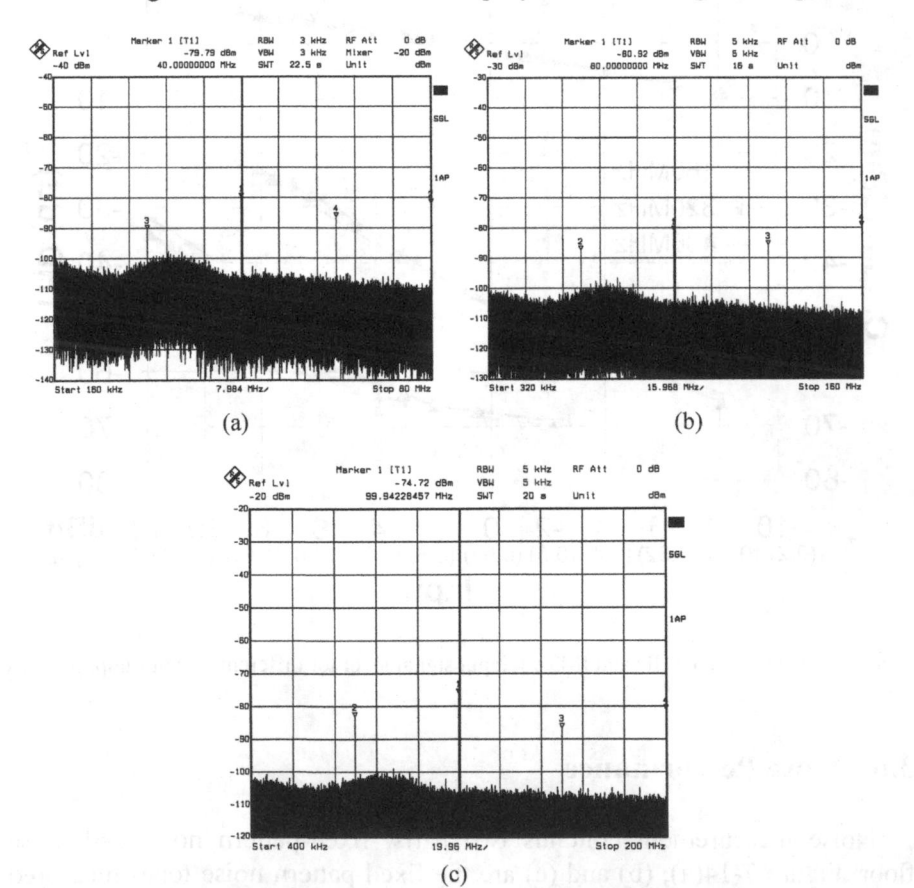

(a)   (b)

(c)

Figure 7-14. Measured fixed-pattern noise with zero input for (a) 160 MHz (b) 320 MHz (c) 400 MHz output sampling rates

Figure 7-15 presents the measured output noise spectrum density for 3 different sampling rates obtained with zero input (including the T/H and output driver that only degrades performance). The total output noise within the Nyquist band is 262 μV_{rms}, 280 μV_{rms} and 265 μV_{rms} for 160 MHz, 320 MHz and 400 MHz sampling rate, respectively. Note that the total noise for 400 MHz rate is smaller than that for 320 MHz, the reason is that, due to the boosted digital supply, the on-resistance of the switches for the former case is much smaller than the latter one, and as analyzed in Chapter 6 / Session 3.6, the on-resistance of all the switches in the charge transferring path in the signal output phase affects the total noise contribution, and the measured results are quite consistent with the estimated noise presented there. According to the measured THD and IM3 results shown in previous sessions 3.3 and 3.4, the corresponding dynamic ranges for 1 % THD & 1% IM3 are 69.4 dB & 64 dB, 68.5 dB & 61dB and 68 dB & 62 dB respectively for 160 MHz, 320 MHz and 400 MHz sampling rates. Moreover, for 1 V_{p-p} input, the corresponding Signal-to-Noise-Plus-Distortion (SINAD), counting the THD, total noise, total pattern noise, as well as mismatch-modulation tones, for the 3 different sampling rates also, are 63 dB, 61 dB and 56 dB, respectively.

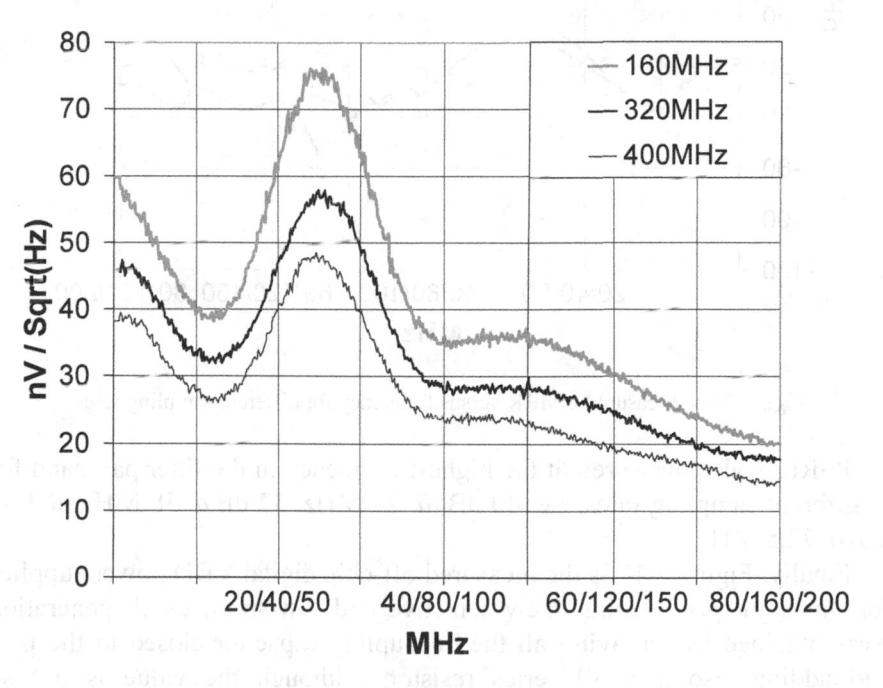

Figure 7-15. Measured output noise spectrum density for different sampling rates

3.7 CMRR and PSRR

To evaluate the circuit immunity to the common mode signals, the measured CMRR versus frequency is presented in Figure 7-16 where it is shown that the CM rejection is always greater than 53 dB for 3 different sampling rates. The single-tone measurement further verifies that the CM rejection is about 56 dB, 55 dB and 53 dB, respectively, by inputting a CM signal at 29 MHz, 58 MHz and 72.5 MHz for 160 MHz, 320 MHz and 400 MHz sampling rates.

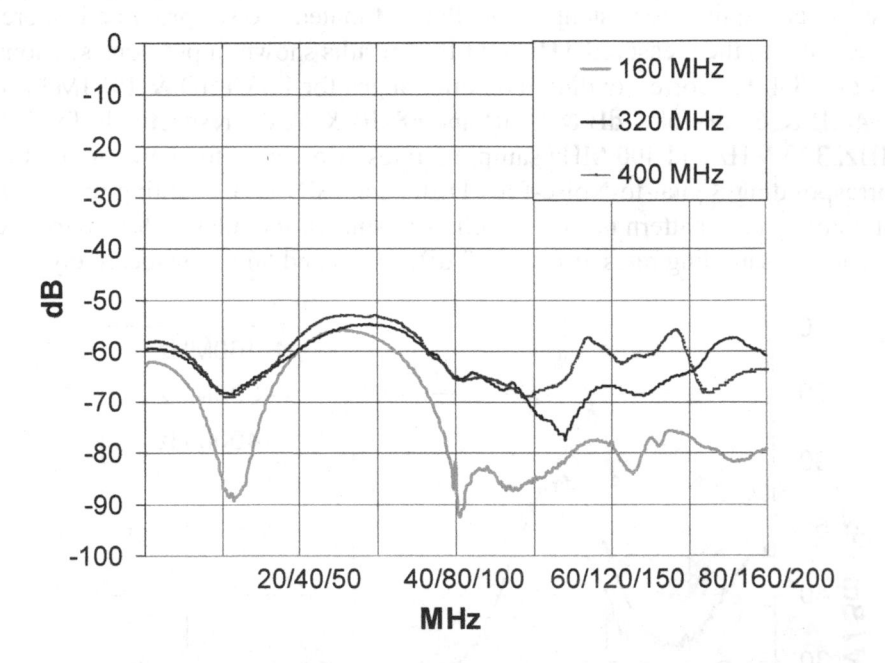

Figure 7-16. Measured CMRR versus frequency for different sampling rates

PSRR is also measured at the highest frequency in the filter passband for 3 different sampling rates, i.e. 40 dB @ 29 MHz, 32 dB@ 58 MHz and 38 dB @ 72.5 MHz.

Finally, Figure 7-17 is the measured off-chip digital VDD power supplies for the high-speed timing-skew sensitive and low-speed clock generation parts obtained by removing all the decoupling capacitor closed to the pins and adding also a small series resistor. Although the value is not so meaningful, it shows clearly the unbalance between the two part VDD supplies and also that the coupling between them is well minimized.

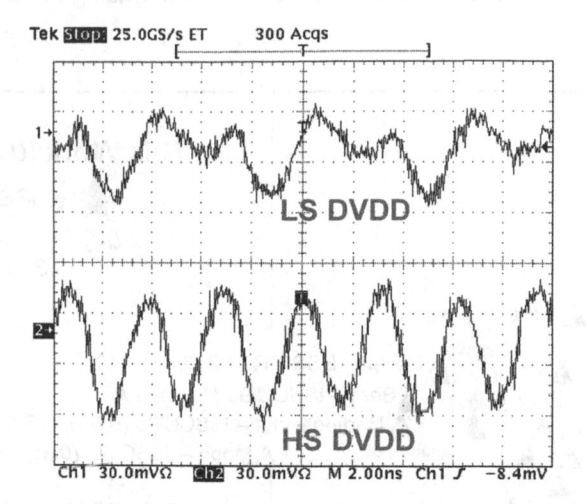

Figure 7-17. Measured off-chip digital power supplies (DVDD=2.5V)

The total analog and digital power occupied by the filter excluding the input T/H and output buffer stages for 160 MHz, 320 MHz and 400 MHz sampling rate are 67, 136, 157mW, corresponding to 4 mW (160 MHz) or 8 mW (320/400 MHz) analog power per tap.

4. SUMMARY

This chapter has presented the measurements for the first/ever designed high-frequency SC bandpass filter prototype that embeds simultaneously the sampling rate increase and a frequency up-translation operation in monolithic CMOS [7.18, 7.19]. The presentation includes the design considerations for the low EMC PCB and testing setup. Comprehensive measurement results have verified that the implemented high-frequency high-order SC filter operates effectively with high dynamic range, high linearity, low fixed-pattern noise, as well as, low mismatch-modulated effects due to the phase timing-skew and parallel-path mismatches. The overall performance of the filter, together with a detailed comparison with the previously reported CMOS SC filters [7.20, 7.21, 7.22, 7.23, 7.24, 7.25, 7.26], in recent 10 years that are within 1 decade lower than the output sampling rate reported in this work have been summarized in the Table 7-3 and also the Figure 7-18. The comparison results show that the current design achieves highest output sampling frequency, highest filter order with highest center frequency and highest dynamic range associated also with the lowest supply voltage.

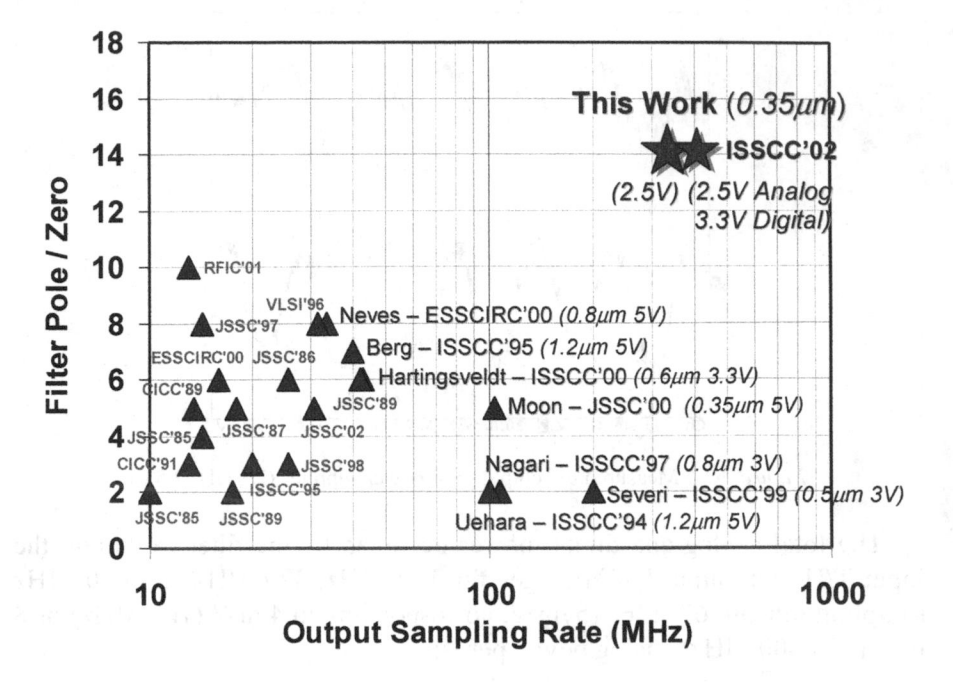

Figure 7-18. Brief comparison of the state-of-the-art CMOS SC filters

Table 7-3. Performance summary of the prototype SC filter with also a comparison with the state-of-the-art CMOS SC filters

	This Design ISSCC'02 [7.8, 7.19]		Severi ISSCC'99 [7.24] JSSC'00	Nagari ISSCC'97 [7.23] JSSC'98	Moon JSSC'10 [7.25]	Uehara ISSCC'94 [7.20] JSSC'94	Quinn ISSCC'00 [7.22] JSSC'00	Berg ISSCC'95 [7.21] JSSC'95	Neves ESSCIRC'00 [7.26]
Technology	0.35 μm CMOS		0.5 μm CMOS	0.8 μm CMOS	0.35 μm CMOS	1.2 μm CMOS	0.6 μm CMOS	1.2 μm CMOS	0.8 μm CMOS
Filter order	15-tap FIR (SC BPF)		2nd-order (SC LPF)	2nd-order (SC BPF)	5th-order (Programmable SC LPF)	3-tap FIR (SC LPF)+3-tap (SC EQ)	6th-order IIR (Variable Gain SC BPF)	5th-order IIR + 3-tap FIR (SC LPF)	9-tap FIR (SC BPF)
Output Sampling Rate	400 MHz	320 MHz	160 MHz	200 MHz	107 MHz	103.68 MHz	100 MHz	40 MHz	33.3 MHz
Input Sampling Rate	100 MHz	80 MHz	40 MHz	200 MHz	107 MHz	300 MHz	100 MHz	80 MHz	33.3 MHz
Passband	f_0=71.25 MHz	f_0=57 MHz	f_{3dB}=20 MHz	f_0=10.7 MHz, Q = 10	f_0=8,12,20 MHz	Notch at 150 MHz	f_0=10.7 MHz, Q = 55		f_0=37.5 MHz
Stopband Rejection	> 40 dB	> 45 dB	> 50 dB	< 35 dB	40 dB			40 dB	40 dB
THD	-57 dB, 1 V_{p-p}, f_{out}=72.5 MHz	-62 dB, 1 V_{p-p}, f_{out}=58 MHz	-64 dB, 1 V_{p-p}, f_{out}=29 MHz	-48 dB, 1 V_{p-p}, f_{out}=4 MHz	-60 dB, 2 V_{p-p}, f_{out}=4 MHz	6-bit Accuracy	-66 dBc, 2 V_{p-p}, f_{out}=100 KHz	-66 dB, 2 V_{p-p}, f_{out}=100 KHz	-44 dB, 1.3 V_{p-p}, f_{out}=16.6 MHz
IM3	-53 dB, f_{out}=72.5 MHz	-52 dB, f_{out}=58 MHz	-62 dB, f_{out}=29 MHz	-62 dB, f_{out}=4 MHz	-54 dB		< -66 dBc, f_{out}=100 KHz	-49 dB, f_{out}=100 KHz	-30 dBc, f_{out}=16.6 MHz
Fixed-Pattern Noise Tones	< -65 dBc	< -70 dBc	< -70 dBc	< -70 dBc		< -66 dBc	< -65 dBc	< -55 dBc	< -30 dBc
Mismatch-Modulated Tones	<-70 dBc	<-72 dBc	<-77 dBc	<-70 dBc					
Total Output Noise	265 μV_{rms}	280 μV_{rms}	560 μV_{rms}	707 μV_{rms}		226 μV_{rms}		1.2 mV_{rms}	14 mV_{rms}
Total Pattern Noise	0.24 mV_{rms}	0.12 mV_{rms}	0.12 mV_{rms}	1.4 mV_{rms}	1.4 mV_{rms}	1.2 mV_{rms}	-		
SINAD	56 dB, 1 V_{p-p}, f_{out}=72.5 MHz	61 dB, 1 V_{p-p}, f_{out}=58 MHz	63 dB, 1 V_{p-p}, f_{out}=29 MHz						32 dB, 1 V_{p-p}, f_{out}=16.6 MHz
Dynamic Range (1% THD)	68 dB	68.5 dB	69.4 dB	62 dB			58 dB	55 dB	
Dynamic Range (1% IM3)	62 dB	61 dB	64 dB	58.4 dB					
OIP3	22.5 dBm	23 dBm	26.4 dBm	26.4 dBm					
CMRR	53 dB (72.5 MHz)	54.5 dB (58 MHz)	56 dB (29 MHz)	28 dB (10.7 MHz)	55 dB (passband)		74 dB (100 KHz)		
PSRR	38 dB (72.5 MHz)	32 dB (58 MHz)	40 dB (29 MHz)	47 dB (10.7 MHz)	29 dB (passband)		55 dB (100 KHz)	55 dB (100 KHz)	
Supply	2.5 V, 3.3 V (Dig)	2.5 V	3 V	3 V	5 V	5 V	3.3 V	5 V	5 V
Analog Power	120 mW	120 mW	59 mW	10 mW	125 mW	97 mW	11.9 mW	190 mW	125 mW
Analog Power-per-tap/pole	8 mW	8 mW	4 mW	5 mW	25 mW	16.1 mW	2 mW	23.75 mW	13.9 mW
Digital Power	37 mW	15.8 mW	7.8 mW			73 mW	4.2 mW	Off-chip	Off-chip
Total Power	157 mW	136 mW	67 mW	23 mW		170 mW	16.1 mW	16.1 mW	
Active Core Area	2 mm² (BPF+T/H+CLK, 0.1 mm² for CLK)		0.054 mm²	0.3 mm²	0.7 mm²	≈ 9 mm²	0.69 mm²	11 mm²	7.27 mm²

Note 1: THD measured with 20 tones by counting up to 5th harmonic of both input and output signals as well as their folded images by the multirate sampling within Nyquist band.
Note 2: The distortion and noise measurements include the input T/H and output driver.
Note 3: The SINAD measured by counting THD, total output noise, total pattern noise as well as mismatched-modulation tones.
Note 4: Analog power excluding input T/H and output driver.

REFERENCES

[7.1] M.K.Armstrong, "PCB design techniques for lowest-cost EMC compliance: Part 1," *Electronics & Communication Engineering Journal*, pp.185-194, Aug.1999.

[7.2] M.K.Armstrong, "PCB design techniques for lowest-cost EMC compliance: Part 1," *Electronics & Communication Engineering Journal*, pp.218-226, Oct.1999.

[7.3] M.Montrose, *EMC and the printed circuit board: design, theory, and layout made simple*, IEEE Press, 1999.

[7.4] K.Fowler, "Grounding and shielding, Part 2 – Grounding and return," *IEEE Instrument & Measurement Magazine*, pp.45-48, Jun.2000.

[7.5] K.Fowler, "Grounding and shielding, Part 1 – Noise," *IEEE Instrument & Measurement Magazine*, pp.41-44, Jun.2000.

[7.6] Yamaichi Electronics, *IC 198 Series (SMT) Socket data sheet: Quad Flat Package (QFP) - 44 pins*.

[7.7] National Semiconductor, *LM1117/LM1117I 800mA Low-Dropout Linear Regulator Data Sheet*, Oct.2000.

[7.8] Texas Instrument, *The bypass capacitor in high-speed environments*, Application note, Nov.1996.

[7.9] T.H.Hubing, J.L.Drewniak, T.P.Van Doren, D.M.Hockanson, "Power bus decoupling on multilayer printed circuit boards," *IEEE Trans. on Electromagnetic Compatibility*, Vol.37, pp.155-166, May 1995

[7.10] J.Chen, M.Xu, T.H.Hubing, J.L.Drewniak, T.P.Van Doren, R.E.DuBroff, "Experimental evaluation of power bus decoupling on a 4-layer printed circuit board," *Proc. International Symposium on Electromagnetic Compatibility*, Vol.1, 2000.

[7.11] J.K.im, B.Choi, H.Kim, W.Ryu, Y.H.Yun, S.H.Ham, S.H.Kim, Y.H.Lee, J.H.Kim, "Separated role of on-chip and on-PCB decoupling capacitors for reduction of radiated emission on printed circuit board," in *Proc. International Symposium on Electromagnetic Compatibility*, Vol.1, pp.531-536, 2001.

[7.12] National Semiconductor, *LM134/LM234/LM334:3-Terminal Adjustable Current Sources Datasheet*, Mar.2000.

[7.13] Mini-Circuits, Product Catalog, http://www.minicurcyuts.com.

[7.14] S.S.Bedair, I.Wolff, "Fast, accurate and simple approximate analytic formulas for calculating the parameters of supported coplanar waveguides for (M)MIC'S," *IEEE Trans. Microwave Theory & Techniques*, Vol. 40, No. 1, pp.41-48, Jan. 1992.

[7.15] I.J.Bahl, P.Bhartia, *Microwave solid state circuit design*, John Wiley and Sons, Apr. 1998.

[7.16] G.Ghione, C.Naldi, "Parameters of coplanar waveguides with lower ground plane, " Electron. Lett., vol. 19, pp. 734-735, 1983.

[7.17] H.Shigesawa, M.Tsuji, A.A.Oliner, "Conductor-backed slotline and coplanar waveguides: dangers and full wave analysis," *IEEE International Microwave Symposium Digest (MTT-S)*, pp.199-202, 1988.

[7.18] Seng-Pan U, R.P.Martins, J.E.Franca, "A 2.5 V, 57 MHz, 15-Tap SC bandpass interpolating filter with 320 MHz output sampling rate in 0.35μm CMOS," in *ISSCC Digest of Technical Papers*, Vol.45, pp380-381, San Francisco, USA, Feb. 2002.

[7.19] Seng-Pan U, R.P.Martins and J.E.Franca, "A 2.5V 57MHz 15-Tap SC Bandpass Interpolating Filter with 320MHz Output Sampling Rate in 0.35mm CMOS," *IEEE J. of Solid-State Circuits*, pp. 87-99, vol.39, January, 2004.

[7.20] G.T.Uehara, P.R.Gray, "A 100MHz output rate analog-to-digital interface for PRML magnetic-disk read channels in 1.2µm CMOS," in *ISSCC Digest Technical Papers*, pp.280-281, Feb.1994.

[7.21] S.K.Berg, P.J.Hurst, S.H.Lewis, P.T.Wong, "A Switched-Capacitor filter in 2µm CMOS using parallelism to sample at 80MHz," in *ISSCC Dig. Tech. Papers*, pp.62-63, Feb.1994.

[7.22] K.V.Hartingsveldt, P.Quinn, A.V.Roermund, "A. 10.7MHz CMOS SC Radio IF Filter with Variable Gain and a Q of 55," in *ISSCC Digest Technical Papers*, pp152-153, Feb.2000.Philips Semiconductors, "SAA7199B, Digital Video Encoder (DENC) Data Sheet," 1996.

[7.23] A.Nagari, G.Nicollini, "A 3 V 10 MHz pseudo-differential SC bandpass filter using gain enhancement replica amplifier," in *ISSCC Dig. Tech. Papers*, pp.52-53, Feb.1997.

[7.24] F.Severi, A.Baschirotto, R.Castello, "A 200Msample/s 10mW Switched-Capacitor Filter in 0.5µm CMOS Technology" in *ISSCC Digest Technical Papers*, pp.400-401, Feb.1999.

[7.25] U.K.Moon, "CMOS High-Frequency Switched-Capacitor filters for telecommunication applications," *IEEE J. Solid-State Circuits*, vol.35, No.2, pp.212-219, Feb. 2000.

[7.26] R.F.Neves, J.E.Franca, "A CMOS Switched-Capacitor bandpass filter with 100 MSample/s input sampling and frequency downconversion," in *Proc. European Solid-State Circuits Conference (ESSCIRC)*, pp. 248-251, Sep.2000.

Chapter 8
CONCLUSIONS

The research work presented in this book led to the development of new analog interpolation techniques for the implementation of optimum-class multirate sampled-data filters. The efficiency of such techniques was fully demonstrated by the realization in the CMOS technology of two Switch-Capacitor interpolating filters for very high-frequency analog front-end applications. Such filters alleviate the operating speed of the digital signal processing core and also relax the requirements of the digital-to-analog conversion interface, as well as, simultaneously, simplify the post continuous-time smoothing filters, thus rendering lower cost integrated solutions. These novel improved multirate SC polyphase structures allow the operation of the interpolating filter core at the lower input sampling rate and are also immune to the traditional lower-rate sample-and-hold shaping distortion, hence realizing an optimum-class analog interpolation and also showing their great potential for pushing analog front-end filtering to a top-speed envelope.

In Chapter 2, the mathematical characterization on the conventional sampled-data analog interpolation whose response is shaped by undesired input lower-rate S/H effect has been first analyzed. Then, the proposals of the ideal improved analog interpolation model and its traditional bi-phase SC structure implementation that are able to eliminate such S/H distortion have been described. Employing multirate polyphase structures to achieve such improved analog interpolation has proved to be a more practical solution to obtain efficient circuit architectures. To achieve an optimum-class realization in terms of its efficiency in power and silicon consumption, different SC circuit architectures have been subsequently investigated with both FIR and IIR transfer functions, respectively, for low and high selectivity filtering on the basis of the improved multirate ADB polyphase structures.

In Chapter 3, design challenges for the practical implementation in silicon of the aforementioned circuit architectures have been studied comprehensively concerning the imperfections of integrated circuit technology. A detailed analysis has also been derived: a simple power dissipation estimation scheme; expected filter response gain errors with respect to the capacitance ratio mismatches; expected signal-to-noise ratio with respect to both the input-referred DC offset of opamps and the clock phase fixed timing-skew and random jitter with holding effects; and estimated total noise power for the polyphase-based interpolating filter circuits. All of these aspects of design are the important keys to open the doors of high-performance analog system response at very high frequency, having also into consideration the random variations of the physical process.

In Chapter 4, a number of mismatch-free SC delay cells and SC summing circuits have been proposed with the employment of Autozeroing and Correlated-Double Sampling in order to improve circuit sensitivity to the input-referred DC offset and finite-gain of opamps. Different kinds of CDS techniques including CS/H-CDS, EC/H-CDS, CS/P-CDS and EC/P-CDS have been applied in those circuits for narrow-band or wideband gain compensation. Both rigorous mathematic expressions of the gain, phase and offset errors for the proposed SC circuits have been derived together with their verification through computer simulations. From the simulation, it is concluded that the stray capacitors, especially the opamp input node capacitance, lead to different-level degradation of the gain and offset compensation performance. Design examples for a 4th-order Elliptic IIR and a 15-tap lowpass SC interpolating filters have been performed using the proposed building blocks, which show the effectiveness of the gain enhancement. The AC analysis for SC CDS circuits has also been presented. Due to the increased total effective capacitive loading, the EC/CDS needs to consume higher power for achieving better performance when compared with the CS/CDS in terms of compensation accuracy, as well as flexible output phase arrangement.

Chapter 5 has presented a design example of an 8-fold 108 MHz output multistage SC linear-phase FIR interpolating filter for NTSC/PAL digital video according to CCIR 601 recommendations. Tailor-made design procedures based on the structures investigated in previous chapters using 0.35 μm CMOS technology, have been described in detail for an optimum and application-specific implementation, which is the rule-of-thumb for high-performance and high-frequency multirate circuit design. This implies multistage, half-band ADB polyphase structure, mismatch-free and multi-unit ADB Semi-Autozeroing scheme, novel coefficient-sharing and two-step summing techniques as well as double-sampling. The performance of the design has also been illustrated through frequency- and time-domain behavioral-, transistor- and parasitic-involved, layout-extracted level simulations. The filter, including both the analog and digital parts, consumes 3.3 mm^2 active area, less than 50 mW static analog and 30 mW average digital power at 3V supply.

In Chapter 6, the prototype specific and optimum design and implementation of a 320 MHz SC bandpass interpolating filter with 15-tap FIR and 57 MHz center frequency for DDFS systems have been presented to up-translate 22-24 MHz inputs sampled at 80 MHz to the 56-58 MHz output band with 4-fold sampling rate increase to 320 MHz. Different design challenges arisen from not only the architectural/circuit-level, but also the layout considerations have been dealt with: including the high-order filtering function, coefficient-sensitivity effects, long power-consuming error-accumulating analog delay line, high-speed output multiplexer and opamp, fixed pattern-noise disturbance, parallel-path and phase timing mismatch-modulated noise, substrate and dI/dt supply noise coupling. Those design techniques have been thoroughly investigated and addressed with thorough verification from behavioral, transistor-level and post layout-extracted CAD simulations, considering the worst-case process variations.

Chapter 7 has presented the experimental verifications of the developed filter prototype fabricated in 0.35 μm double-poly triple-metal CMOS technology. The low EMC PCB design techniques for the performance evaluation of this high-frequency prototype and the corresponding testing setup have been firstly addressed. The experimental verification has then been performed thoroughly at different sampling rates with respect to the frequency-domain measurement, e.g. amplitude response, group delay, THD & IM3 versus the input of different signal levels, noise, CMRR, and others, as well as time-domain functional measurements. The measurement results have shown that the filter operates with full functional capability and excellent consistence with the theoretical expectations – even better than the worst-case simulations, thus consolidating the effectiveness of all the design techniques developed which can be replicated into any other high-frequency SC filter implementations.

The prototype SC filter that embeds for the first-time in an IC design the sampling rate increase and frequency up-translation operation operates simultaneously at the nominal 320 MHz and also at 400 MHz, still satisfying the design specifications. It occupies 2 mm^2 active area, 120 mW for analog and 15.8 mW for digital power corresponding to 8 mW per zero at 2.5 V supply for nominal 320 MHz rate and achieves high linearity (62 dB THD, 52 dB IM3), low noise (280 μV$_{rms}$) and thus high dynamic range (69 dB for

1% THD, 61 dB for 1% IM3, 61 dB SINAD), low fixed-pattern noise (<-70 dBc), as well as low parallel-path and timing-skew mismatch-modulated effects (<-72 dBc). The detailed comparison with previously reported CMOS SC filters, in the last 10 years in recent 10 years that are within 1 decade lower than the output sampling rate reported in this work, demonstrates that the current design operates at the highest sampling rate and achieves highest filter order and center frequency with the highest dynamic range and the lowest supply voltage.

In summary, from the silicon results which match splendidly with theoretical expectations, the effectiveness of the presented system/architectural-, circuit- and layout-level optimization schemes wrestling with different design challenges at such high-frequency operation are validated for state-of-the-art high-speed analog frond-end filtering.

Appendix 1

TIMING-MISMATCH ERRORS WITH NONUNIFORMLY HOLDING EFFECTS

Due to the inherent different nature of signal sampling and processing in the applications, the following 3 different types of processes can represent timing-mismatch effects:

TYPE 1: If the **I**nput signal is sampled by the system with **N**onuniform time-interval and later played out or represented by discrete samples in the **O**utput at **U**niform time instants for later processing, it can be designated by an **IN-OU** process. This is a typical sampling process in the analog to digital conversion path (timing-mismatch only @ input signal sampling) of TI ADCs [1.47] or multirate sampled-data analog decimators [1.51], as shown in Figure A1-1(a) with the correspondent signal waveform.

TYPE 2: If the **I**nput signal is sampled by the system with **U**niformly spaced time-intervals and later the samples are played out in the **O**utput **N**onuniformly, then the system is referred to as **IU-ON** process, that is the typical case in digital to analog conversion path (timing-mismatch only @ output signal holding) of TI DACs [3.20] or multirate sampled-data analog interpolators [6.55, 6.56], as shown in Figure A1-1(b).

TYPE 3:. If the **I**nput signal is sampled by the system with **N**onuniform time-interval and then played out **C**orrelatively at the **O**utput with the same **N**onuniform time instants (occurred at the input sampling), such system is named **IN-CON** process. It can typically be applied to a complete TI sampled-data system (timing-mismatches @ both input sampling and output holding driven by the same clock phases), as illustrated in Figure A1-1(c), e.g. N-path filtering [1.59].

The signal spectra for these 3 types of processes with Impulse-Sampled (IS) sequence form have been well analyzed in [3.6,3.13,3.14] respectively. However, in practice, the real output signals are always in Sample-and-Hold (SH) or holding nature in the latter two processes, as represented in the waveforms of Figure A1-1. Figure A1-2 presents the plots of FFT output spectra of all IN-OU, IU-ON and IN-CON process with both IS and SH output for a sinusoidal input with normalized frequency $a=f_o/f_s=0.2$, timing-skew period $M=8$ and standard deviation of timing-skew ratio $\sigma_{rm} = 0.1\%$, where showing clearly the modulated imaging sidebands are located around f_s/M and its multiples for all three cases. Unlike the case of IN-OU process, the output signal spectra of the latter two processes are not simply the $\sin(x)/x$ shaped version of the corresponding original impulse-sampled output due to the nonuniformly holding effect, e.g. compare the calculated SINAD and circled parts in the figure for all cases (the sideband magnitudes of SH version of the IN-OU output signal decrease gradually as frequency increases due to the typical uniformly zero-order hold transfer function, while for the latter two cases there are nonuniform modifications for all the sidebands). Therefore, the previous analysis of IU-ON(IS) [3.14] and IN-CON(IS) [3.13] cannot be directly applicable for their corresponding SH versions. This appendix will present a complete investigation of output signal spectra of practical IU-ON(SH) and IN-CON(SH) processes, including both the closed-form spectra expression as well as their output SINAD. Furthermore, the Spurious Free Dynamic Range (SFDR) subjected to in-band image tone will also be derived for a main practical application of the IN-CON(SH) process, i.e. N-path filtering. The corresponding simulation results for demonstration of the accuracy of the derived formula will then also be presented.

Besides, by carefully investigation among the spectrum of three cases, as shown in Figure A1-2, it can be found that the spectrum of IN-OU(IS) and IU-ON(SH) exhibit great similarity under the assumption of $2\pi f_o r_m T <<1$: not only similar from the plot but also their SINAD and SFDR within the Nyquist band are all the same. The second part of this appendix will prove the correlation between them and show that the noise components for these two cases are exactly the same.

1. SPECTRUM EXPRESSIONS FOR IU-ON(SH) AND IN-CON(SH)

1.1 IU-ON(SH)

Consider the IU-ON(SH), i.e. the system ideally plays out the uniformly sampled input samples at nonuniformly spaced time intervals with holding, i.e.

$$y_{IU-ON(SH)}(t) = \sum_{n=-\infty}^{\infty} x(nT)h_n(t-t_n) \tag{A1.1}$$

where

$$h_n(t) = u(t) - u(t - T - \Delta_{n+1} + \Delta_n) \tag{A1.1a}$$

$$t_n = nT + \Delta_n \tag{A1.1b}$$

and T is the nominal sampling period and Δ_n is a periodic skew timing sequence with period M (path number). Let $n=kM+m$ (m‒0,1,…M-1) and $r_m = \Delta_m / T$, then we have

$$y_{IU-ON(SH)}(t) = \sum_{m=0}^{M-1} \sum_{k=-\infty}^{\infty} x(kMT + mT)h_m(t - kMT - mT - r_mT) \tag{A1.2}$$

Appling the Fourier Transform to $y_{IU\cdot ON(SH)}(t)$, we have

$$Y_{IU-ON(SH)}(\omega) = \sum_{m=0}^{M-1} H_m(\omega) \left(\sum_{k=-\infty}^{\infty} x(kMT + mT)e^{-j\omega kMT}e^{-j\omega mT}e^{-j\omega r_mT} \right) \tag{A1.3}$$

and by substituting the following identity

$$x(kMT + mT) = \frac{1}{2\pi} \int_{-\infty}^{\infty} X(\Omega)e^{j\Omega(kMT+mT)} d\Omega \tag{A1.4}$$

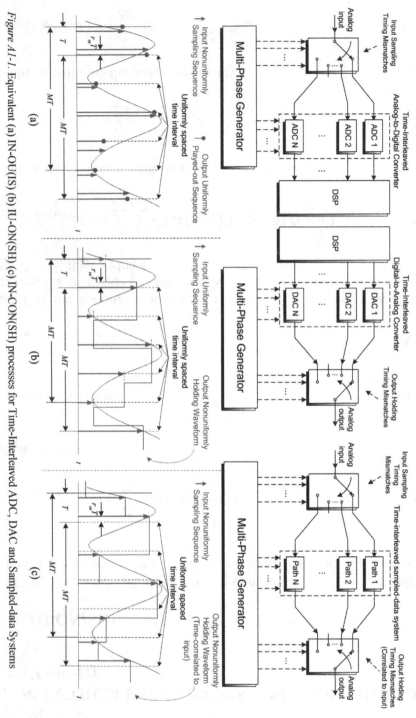

Figure A1-1. Equivalent (a) IN-OU(IS) (b) IU-ON(SH) (c) IN-CON(SH) processes for Time-Interleaved ADC, DAC and Sampled-data Systems

Design of Very High-Frequency Multirate Switched-Capacitor Circuits –
Extending the Boundaries of CMOS Analog Front-End Filtering

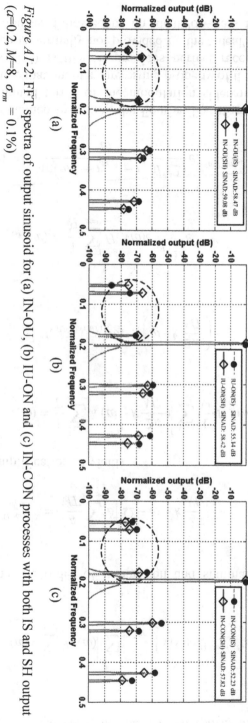

Figure A1-2: FFT spectra of output sinusoid for (a) IN-OU, (b) IU-ON and (c) IN-CON processes with both IS and SH output (a=0.2, M=8, σ_{rm} = 0.1%)

we obtain

$$Y_{IU-ON(SH)}(\omega) = \sum_{m=0}^{M-1} H_m(\omega) \left[\sum_{k=-\infty}^{\infty} \left(\frac{1}{2\pi} \int_{-\infty}^{\infty} X(\Omega) e^{j\Omega(kMT+mT)} d\Omega \right) e^{-j\omega kMT} e^{-j\omega mT} e^{-j\omega r_m T} \right]$$

(A1.5)

By changing the order of integration and summation, and applying the following:

$$\sum_{k=-\infty}^{\infty} e^{jk(\Omega-\omega)MT} = \frac{2\pi}{MT} \sum_{k=-\infty}^{\infty} \delta\left(\Omega - \omega + k\frac{2\pi}{MT} \right)$$

(A1.6)

and after simplification, it yields

$$Y_{IU-ON(SH)}(\omega) = \frac{1}{T} \sum_{k=-\infty}^{\infty} X(\omega - k\frac{2\pi}{MT}) \left[\sum_{m=0}^{M-1} \frac{1}{M} H_m(\omega) e^{-jkm\frac{2\pi}{M}} e^{-j\omega r_m T} \right]$$

$$= \frac{1}{T} \sum_{k=-\infty}^{\infty} A_{k,IU-ON(SH)}(\omega) \cdot X\left(\omega - k\frac{2\pi}{MT} \right)$$

(A1.7)

where

$$A_{k,IU-ON(SH)}(\omega) = \frac{1}{M} \sum_{m=0}^{M-1} H_m(\omega) e^{-jkm\frac{2\pi}{M}} e^{-j\omega r_m T}$$

(A1.7a)

$$H_m(\omega) = \frac{2\sin\left(\omega(1 + r_{m+1} - r_m)T/2\right)}{\omega} e^{-j\omega(1+r_{m+1}-r_m)T/2}$$

(A1.7b)

(A1.7b) is obtained by taking the Fourier Transform of (A1.1a). $A_{k,IU-ON(SH)}$ are the weighted terms of the modulation sidebands for the IU-ON(SH) system. Notice that another interesting property of the processes described in (A1.1) to (A1.7) is that, it provides perfect reconstruction of DC-level signals which is normally not provided by the "ideal" reconstruction method: weighted impulses followed by an ideal low-pass filter. Furthermore, DC

constant does not have the error caused by nonunifomly play-out time
instance imposed by (A1.1a).

1.2 IN-CON(SH)

For the IN-CON(SH), where the output is described by

$$y_{IN-CON(SH)}(t) = \sum_{n=-\infty}^{\infty} x(t_n)h_n(t-t_n)$$

$$= \sum_{m=0}^{M-1} \sum_{k=-\infty}^{\infty} x(kMT + mT + r_m T)h_m(t - kMT - mT - r_m T) \qquad (A1.8)$$

By the similar procedures, we have

$$Y_{IN-CON(SH)}(\omega) = \frac{1}{T} \sum_{k=-\infty}^{\infty} X(\omega - k\frac{2\pi}{MT}) \left[\sum_{m=0}^{M-1} \frac{1}{M} H_m(\omega)e^{-jkm\frac{2\pi}{M}} e^{-jkr_m\frac{2\pi}{M}} \right]$$

$$= \frac{1}{T} \sum_{k=-\infty}^{\infty} A_{k,IN-CON(SH)}(\omega) \cdot X\left(\omega - k\frac{2\pi}{MT}\right) \qquad (A1.9)$$

where

$$A_{k,IN-CON(SH)}(\omega) = \frac{1}{M} \sum_{m=0}^{M-1} H_m(\omega)e^{-jkr_m\frac{2\pi}{M}} e^{-jkm\frac{2\pi}{M}} \qquad (A1.9a)$$

The equations (A1.7), (A1.7a) (A1.7b) (A1.9) and (A1.9a) fully characterize
the output signal spectrum for IU-ON(SH) and IN-CON(SH) cases.

2. CLOSED FORM SINAD EXPRESSION FOR IU-ON(SH) AND IN-CON(SH)

From the derived sidebands weighted terms in (A1.7a) and (A1.9a), the
modulation sidebands are centered at frequencies of $\omega = \pm\omega_0 + k(2\pi)/(MT)$,

and thus the signal and image distortion components can be both represented by A_k's with $k = 0$ and $k \neq 0$, respectively. Consequently, the expected signal and distortion power of the output can be evaluated separately by statistical means so as to obtain the SINAD. Similarly, the in-band SFDR for applications of IN-CON(SH) systems, e.g. N-path narrow-band filtering, can be accomplished with the evaluation of A_k with the value of k corresponding to in-band image tone.

2.1 IU-ON(SH)

Consider first the IU-ON(SH) system described in (A1.7a) and (A1.7b). For a real input sinusoidal signal with frequency $\omega_o = 2\pi f_o$, the sidebands are located at $\omega = \pm\omega_o + k(2\pi)/(MT)$, with $k=0$ representing signal and $k = +1,+2,...$ or $k = -1,-2,...$ representing images components at positive and negative frequency axis, respectively. Then, it would be possible to evaluate the sideband components at only $\omega = \omega_o + k(2\pi)/(MT)$ over the range $[-f_s/2, f_s/2]$ to find the SINAD over the range of $[0, f_s/2]$, since the images at positive frequencies of $\omega = -\omega_o + k(2\pi)/(MT)$ with $k = +1,+2,...$ will be directly reflected by the images at negative frequencies of $\omega = \omega_o + k(2\pi)/(MT)$ with $k = -1,-2,...$, by the symmetry property of Fourier Transform of real signal.. In the following formula derivation, it is assumed that $\omega_o \neq k(2\pi)/(MT)$, meaning that the signal (and also the sidebands) is not exactly located at integer multiple of f_s/M. Simplifying (A1.7a) and (A1.7b) using $\omega = \omega_o + k(2\pi)/(MT)$ and considering that r_m and $e^{-jkm(2\pi)/M}$ are periodic with period $m=M$, will lead to

$$A_{k,IU-ON(SH)}\left(\omega_o + k\frac{2\pi}{MT}\right) = \frac{2\sin(\omega_o T/2)}{\left(\omega_o + k\dfrac{2\pi}{MT}\right)M} e^{-j\omega_o T/2} \sum_{m=0}^{M-1} e^{-j\omega_o r_m T} e^{-jkr_m\frac{2\pi}{M}} e^{-jkm\frac{2\pi}{M}}$$

(A1.10)

and the SINAD can be found by the following formula:

$$SINAD = 10\log_{10}\left[\frac{\left|A_{0,IU-ON(SH)}(\omega_0)\right|^2}{\displaystyle\sum_k \left|A_{k,IU-ON(SH)}\left(\omega_0 + k\frac{2\pi}{MT}\right)\right|^2}\right] dB$$

(A1.11)

where the value of k in the summation is taken in such a way $-\pi f_s \leq \omega_o + k(2\pi)/(MT) \leq \pi f_s$ and $k \neq 0$. Assume now r_m ($m = 0,1,2,...M\text{-}1$) to be M independent identically distributed (i.i.d) random variables with Gaussian distribution of zero mean and standard deviation σ_{rm} ($= \sigma_t / T$ where σ_t is the standard deviation of timing jitter in second). Thus, the expected values of the signal component $|A_{0,IU-ON(SH)}(\omega_o)|^2$ can be evaluated by substituting $k = 0$ into (A1.10) and multiplying (A1.10) by its complex conjugate as follows:

$$E[|A_{0,IU-ON(SH)}(\omega_0)|^2] = \frac{4}{\omega_0^2 M^2} \sin^2(\omega_0 T/2) \sum_{n=0}^{M-1}\sum_{m=0}^{M-1} E[e^{-j\omega_0(r_m-r_n)T}] \quad (A1.12)$$

When $\omega_0 r_m T \ll 1$ (or equivalently $\omega_0 \sigma_{rm} T \ll 1$) for small values of r_m, (A1.12) can be simplified to

$$E[|A_{0,IU-ON(SH)}(\omega_0)|^2] \approx \frac{4}{\omega_0^2 M^2} \sin^2(\omega_0 T/2) \sum_{n=0}^{M-1}\sum_{m=0}^{M-1} E[1 - j\omega_0(r_m - r_n)T] \approx \frac{4}{\omega_0^2} \sin^2(\omega_0 T/2)$$

$$(A1.13)$$

Similarly, the sideband components $|A_{k,IU-ON(SH)}(\omega_o + 2\pi k/MT)|^2$ can be calculated from (A1.10) as follows:

$$E\left[\left|A_{k,IU-ON(SH)}\left(\omega_0 + k \cdot \frac{2\pi}{MT}\right)\right|^2\right] \approx \frac{4\sigma_{rm}^2 T^2}{M} \sin^2(\omega_0 T/2) \quad (A1.14)$$

From (A1.14) it is clear that the expected value of various distortion components in IU-ON(SH) systems is identical for different values of k, thus the total distortion power can be obtained by simply multiplying (A1.14) by $M\text{-}1$. Finally using (A1.13) and (A1.14) in (A1.11), the SINAD of IU-ON(SH) systems can be obtained as follows:

$$SINAD_{IU-ON(SH)} = 20\log_{10}\left(\frac{1}{2\pi a\sigma_{rm}}\right) - 10\log_{10}\left(1 - \frac{1}{M}\right) \quad (A1.15)$$

where $a = f_o/f_s$ is the normalized frequency of the sinusoid. Note that the first term of (A1.15) describes the pure random clock jitter, which is obtained by letting time-skew sequences period $M \rightarrow \infty$. Moreover, the SINAD formula (A1.15) for IU-ON(SH) is interestingly identical to the that for traditional non-uniformly impulse sampling IN-OU(IS) system [3.6] with small jitter errors. Indeed when $2\pi f_0 \sigma_{rm} T \ll 1$, the normalized sideband spectra patterns for IN-OU(IS) and IN-OU(SH) are proved to be identical in the following session (this is also evident from the spectra and SINAD shown in Figure A1-2).

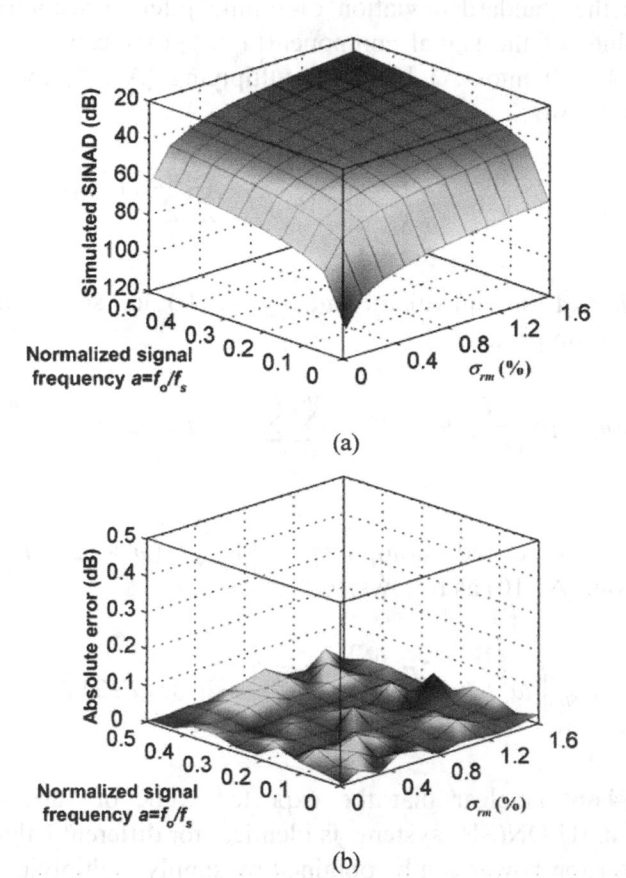

(a)

(b)

Figure A1-3. (a) Simulated SINAD & (b) absolute error between the simulated and calculated SINAD of IU-ON(SH) systems vs. normalized frequency a and standard derivation σ_{rm} by 10^4 times Monte Carlo Simulations (M=8)

Figure A1-3(a) shows the MATLAB simulation results of a IU-ON(SH) system (obtained with M=8) and in Figure A1-3(b) the absolute error between the simulated and the calculated results is presented (against a standard deviation σ_{rm} and normalized signal frequency $a=f_0/f_s$) illustrating the accuracy of the derived formula (A1.15). The error between the IU-ON(SH) output SINAD and that predicted by the formula (A1.15) is well below 0.1dB, as shown in Figure A1-3(b), thus confirming the consistency between the theoretical prediction and the practical simulations.

2.2 IN-CON(SH)

For the IN-CON(SH) system, substituting $\omega = \omega_o + k(2\pi)/(MT)$ in (A1.9) yields:

$$A_{k,IN-CON(SH)}(\omega_0 + k\frac{2\pi}{MT}) = \frac{1}{j(\omega_0 + k\frac{2\pi}{MT})M}\left\{\sum_{m=0}^{M-1} e^{-jkr_m\frac{2\pi}{M}} e^{-jkm\frac{2\pi}{M}} - e^{-j\omega_0 T}\sum_{m=0}^{M-1} e^{-j\omega_0 r_m T} e^{-jkr_m\frac{2\pi}{M}} e^{j\omega_0 r_{m-1} T} e^{-jkm\frac{2\pi}{M}}\right\}$$

(A1.16)

and then, with an expected value of the signal component is identical to (A1.13), the expected value of the distortion components can be expressed as follows:

$$E[\left|A_{k,IN-CON(SH)}(\omega_0 + k\frac{2\pi}{MT})\right|^2] \approx \frac{2\sigma_{rm}^2 T^2}{M}\left\{2\sin^2(\omega_0 T/2) - 2\omega_0 T\sin^2(\omega_0 T/2) \cdot \frac{1-\cos(\omega_0 T + k\frac{2\pi}{M})}{\omega_0 T + k\frac{2\pi}{M}}\right.$$
$$\left. + \omega_0^2 T^2 \cdot \frac{1-\cos(\omega_0 T + k\frac{2\pi}{M})}{\left(\omega_0 T + k\frac{2\pi}{M}\right)^2} - \omega_0 T\sin(\omega_0 T) \cdot \frac{\sin(\omega_0 T + k\frac{2\pi}{M})}{\omega_0 T + k\frac{2\pi}{M}}\right\}$$

(A1.17)

Unlike in the case of IN-CON(SH), the expected distortion components of the IN-CON(SH) system depends on the value of k, thus it is important to determine the expected values of the total power of all noise components at $\omega = \pm\omega_o + k(2\pi)/(MT)$ in the range of $[0, f_s/2]$, and the previous sum in (A1.11) should be modified to include also the images at $\omega = -\omega_o + k(2\pi)/(MT)$, with k satisfying the following inequality:

$$0 < \pm\omega_0 T + k\frac{2\pi}{M} < \pi \text{ and } k \neq 0$$

(A1.18)

The direct calculation of the distortion power sum over this range is not realistic due to the high level of complexity in (A1.17). However, an important characteristic in (A1.17) is that the expected value of the distortion components is a function of $\omega_s T + k(2\pi)/M$, so for the case of purely random timing jitter ($M \rightarrow \infty$), the sum tends to an integral and it is possible to calculate the sum in (A1.11) by integrating (A1.17) over the limit specified in (A1.18). Thus, after the integration, the total noise power at $\omega = \pm\omega_o + k(2\pi)/(MT)$ will be given by:

$$\sum_k E\left[\left|A_{k,IN-CON(SH)}\left(\pm\omega_o + k\frac{2\pi}{MT}\right)\right|^2\right] = 4\sigma_{rm}^2 T^2 \sin^2(\pi a) \cdot [0.5\,\mu\,1.65a - \frac{1.85a}{\tan(\pi a)} + \frac{3.82a^2}{\sin^2(\pi a)}]$$

(A1.19)

Then, by using (A1.11), (A1.13) and (A1.19), the closed-form expression of SINAD for IN-CON(SH) system is obtained by

$$SINAD_{IN-CON(SH)} \approx 20\log_{10}\left(\frac{1}{2\pi a \sigma_{rm}}\right) - 10\log_{10}[1 - \frac{3.7a}{\tan(\pi a)} + \frac{7.64a^2}{\sin^2(\pi a)}]\,(A1.20)$$

Figure A1-4 is the 3D plot showing both the simulated SNR and the prediction error from the derived formula (A1.20) vs. normalized signal frequency a and the standard deviation σ_{rm} for an IN-CON(SH) system. The absolute error is well below 0.2 dB showing again the effectiveness of the derived SINAD formula.

Although the SINAD derived in (A1.20) is appropriate for purely random jitter, it can also approximate the SINAD due to periodic timing-skew. Figure A1-5 shows the approximation error (in dB) between simulated SINAD and the SINAD calculated using (A1.20) as a function of timing-skew period M, normalized signal frequency a and σ_{rm}. Figure A1-5(a) shows that the error converges to zero as the path and thus the periodic timing-skew periodic M becomes large, and even with low values of M ($M=2$ in both Figure A1-5(a) & (b)), the formula can approximate the SNR within approximately 2 dB accuracy range (with 10^3 times Monte Carlo simulations) for all range of signal frequencies. Notice that the error grows as the σ_{rm} becomes large ($\sigma_{rm} > 10\%$), but referred to Figure A1-4(a) the corresponding SNR is already smaller than 20 dB, which proves much less usage for most applications.

3. CLOSED FORM SFDR EXPRESSION FOR IN-CON(SH) SYSTEMS

In addition to the derived SINAD of the IN-CON(SH) system, it is interesting to investigate also the effects of jitter-induced in-band mirror tone in application of IN-CON(SH) system, e.g. N-path sampled-data filtering which is shown in Figure A1-1(c) and is one of the most common methods to construct a narrow-band band-pass filter [1.59,3.24,6.20] with passband centered at f_s/N. However, one of the mirror tones caused by periodic timing-skew always appears in-band with signal as shown in Figure A1-2(c), thus destroying the performance of narrow-band filtering. For the input signal located at a frequency ω_0, the value of k corresponding to the in-band image can be calculated as follows (which is produced by negative frequency components of the signal):

$$\frac{1}{2}\frac{2\pi}{MT} < -\omega_0 + k\frac{2\pi}{MT} < \frac{3}{2}\frac{2\pi}{MT} \text{ or } 0.5 < -Ma + k < 1.5 \tag{A1.21}$$

If the signal is placed close to the center frequency, then $a \approx 1/N = 1/M$ and (A1.21) will give $k = 2$.

The expected value of the mirror tone can be found from (A1.17) with $k = 2$ and $a \approx 1/M$ (using negative frequency components), that is:

$$E[|A_{2,IN-CON(SH)}(-\omega_0 + 2\cdot\frac{2\pi}{MT})|^2] \approx \frac{16\sigma_{rm}^2 T^2}{M}\sin^4\frac{\pi}{M} \tag{A1.22}$$

and the expected value of the signal component results from (A1.13) as:

$$E[|A_{0,IN-CON(SH)}(\omega_0)|^2] \approx \frac{M^2 T^2}{\pi^2}\sin^2\frac{\pi}{M} \tag{A1.23}$$

Thus, from (A1.22) and (A1.23) the Spurious Free Dynamic Range (SFDR) in the passband subjected to jitter is given by:

$$SFDR_{jitter} \approx 30\log_{10} M - 20\log_{10}\left(4\pi\sigma_{rm}\sin\frac{\pi}{M}\right)(dBc) \tag{A1.24}$$

Note that this equation is valid only when $a \approx 1/M$ which is usually true for N-path narrow bandpass filter since the signal is located near the frequency of $f_s/N=f_s/M$. Figure A1-6 shows a plot of (A1.24) as function of the timing skew period M and standard deviation σ_{rm}, and from the figure a considerable reduction on the in-band mirror tone caused by periodic timing-skew is possible via increasing the path number N (and the timing-skew period M).

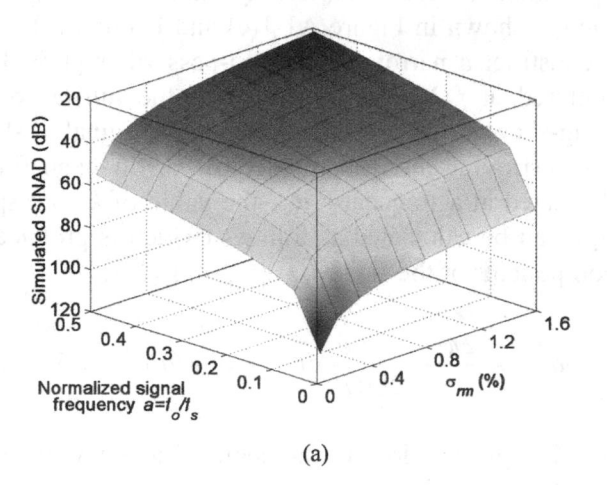

(a)

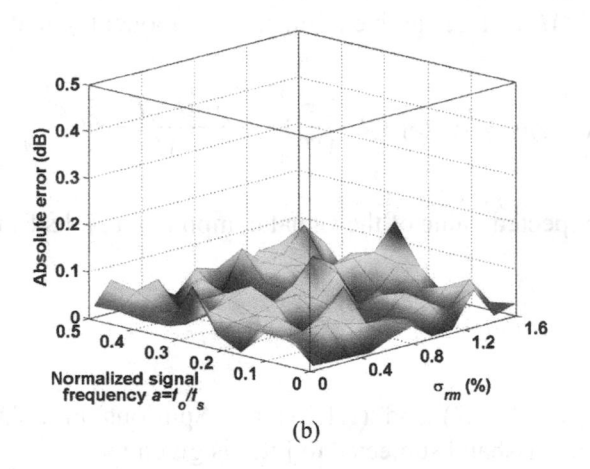

(b)

Figure A1-4. (a) Simulated SINAD & (b) absolute error between the simulated and calculated SINAD of IN-CON(SH) systems vs. normalized frequency a and standard derivation σ_{rm} by 10^3 times Monte Carlo simulations

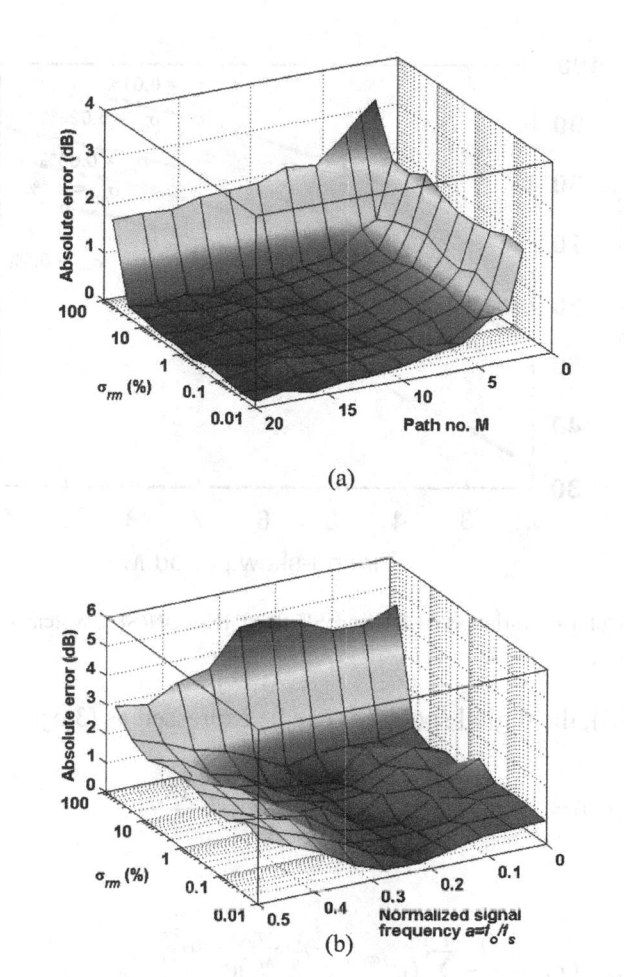

Figure A1-5. Absolute error between the simulated and calculated SINAD of IN-CON(SH) systems vs. (a) path no. M and standard derivation σ_{rm} ($a = 0.5$) and (b) normalized signal frequency a and standard derivation σ_{rm} ($M = 2$) by 10^3 times Monte Carlo simulations

4. SPECTRUM CORRELATION OF IN-OU(IS) AND IU-ON(SH)

Figure A1-7(a) and (b) give more spectrum plot of a 58 MHz signal sampled at 320 MHz with the sampling processes IN-OU(IS) and IU-ON(SH) with $M=4$, $\sigma=20$ ps. Similarly, their normalized noise components (with respect to signal component @ $f_o = 58$ MHz) are equal.

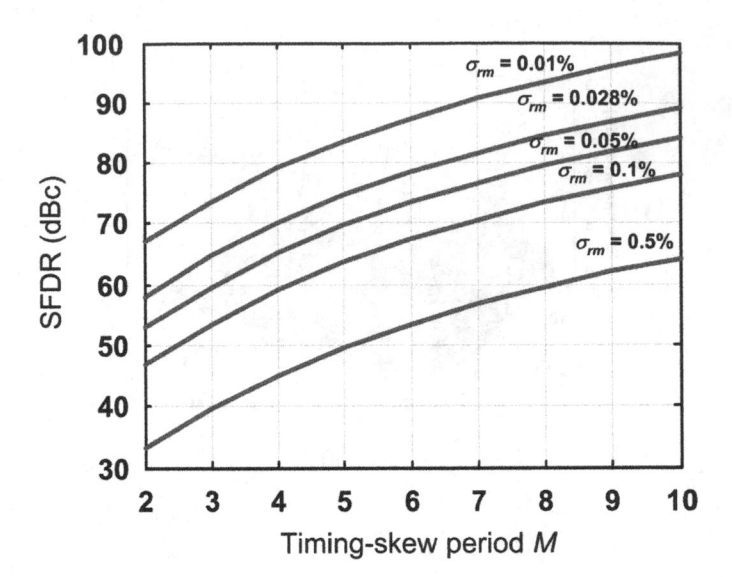

Figure A1-6. A plot of variation of in-band SFDR of IN-CON(SH) system vs. timing-skew
period M and σ_{rm}

For IN-OU(IS), the digital spectrum can be expressed as [3.5]:

$$Y_{IN-OU(IS)}(\omega) = \frac{1}{T} \sum_{k=-\infty}^{\infty} A_{k,IN-OU(IS)}(\omega) \cdot X\left(\omega - k\frac{2\pi}{MT}\right) \tag{A1.25}$$

$$A_{k,IN-OU(IS)}(\omega) = \frac{1}{M} \sum_{m=0}^{M-1} (e^{j\omega r_m T} e^{-jkr_m \frac{2\pi}{M}}) e^{-jkm\frac{2\pi}{M}} \tag{A1.25a}$$

First we consider the distortion produced by the input component
$\pi\delta(\Omega - \Omega_o)$. At frequency location $\omega = \omega_o + k(2\pi)/MT$

$$A_{k,IN-OU(IS)}(\omega_o + k\frac{2\pi}{MT}) = \frac{1}{M} \sum_{m=0}^{M-1} (e^{j\omega_o r_m T}) e^{-jkm\frac{2\pi}{M}} \tag{A1.26}$$

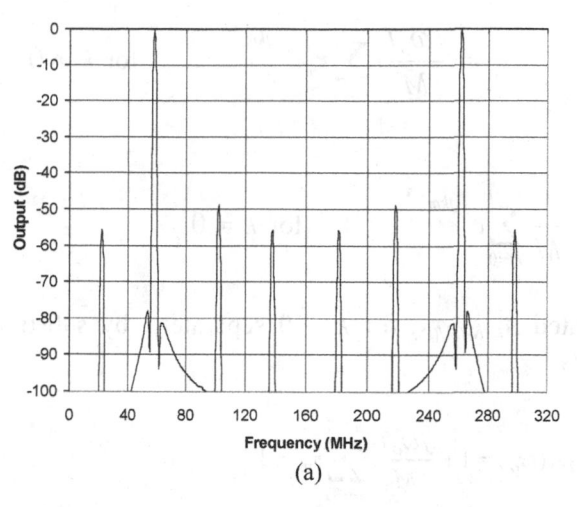

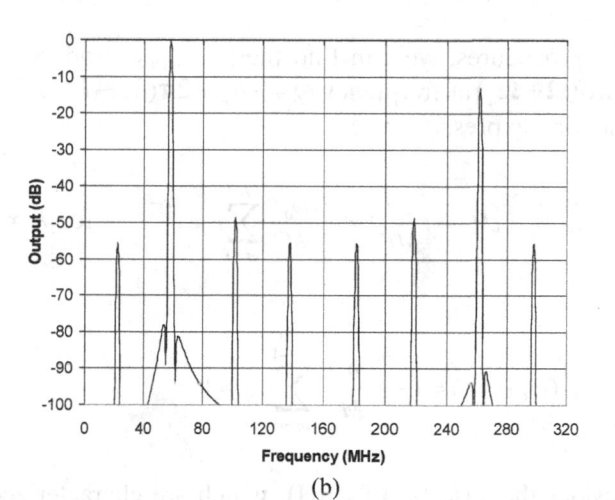

Figure A1-7. FFT of a 58 MHz signal sampled at 320 MHz for (a) IN-OU(IS) (b) IU-ON(SH) $M=4$, $\sigma=20$ ps)

for $2\pi f_o r_m T \ll 1$ and using $e^{j\omega_o r_m T} \approx 1 + j\omega_o r_m T$

$$A_{k,IN-OU(IS)}(\omega_o + k\frac{2\pi}{MT}) \approx \frac{1}{M}\sum_{m=0}^{M-1}(1 + j\omega_o r_m T)e^{-jkm\frac{2\pi}{M}}$$

$$= \frac{1}{M}\sum_{m=0}^{M-1} e^{-jkm\frac{2\pi}{M}} + \frac{j\omega_o T}{M}\sum_{m=0}^{M-1} r_m e^{-jkm\frac{2\pi}{M}} \qquad (A1.27)$$

$$= \frac{j\omega_o T}{M} \sum_{m=0}^{M-1} r_m e^{-jkm\frac{2\pi}{M}} \qquad \text{for } k \neq 0 \qquad (A1.28)$$

Because $\displaystyle\frac{1}{M}\sum_{m=0}^{M-1}e^{-jkm\frac{2\pi}{M}} = 0 \quad \text{for } k \neq 0$

To calculated $A_{k,IN\text{-}OU(IS)}$ for $k = 0$ separately, by substitute $k = 0$ into (A1.27), yields

$$A_{0,IN-OU(IS)}(\omega_o) = 1 + \frac{j\omega_o T}{M} \sum_{m=0}^{M-1} r_m \approx 1 \qquad (A1.29)$$

Using similar procedures, we can find the $A_{k,IN\text{-}OU(IS)}$ produced by the input component $\pi\delta(\Omega + \Omega_o)$ at frequency $\omega = -\omega_o + 2\pi(M-k)/MT$, which will have the following expressions:

$$A_{k,IN-OU(IS)}(-\omega_o + (M-k)\frac{2\pi}{MT}) = -\frac{j\omega_o T}{M} \sum_{m=0}^{M-1} r_m e^{jkm\frac{2\pi}{M}} \quad \text{for } k \neq 0 \quad (A1.30)$$

$$A_{0,IN-OU(IS)}(\omega_s - \omega_0) = 1 - \frac{j\omega_o T}{M} \sum_{m=0}^{M-1} r_m \approx 1 \qquad (A1.31)$$

Now consider the case IU-ON(SH), which are characterized by (A1.7), (A1.7a) and (A1.7b). We denote now the distortion components as $A_{k,IU\text{-}ON(SH)}$ for IU-ON(SH). Substitute (A1.7b) into (A1.7a) and using Euler formula, it yields

$$A_{k,IU-ON(SH)}(\omega) = \frac{1}{j\omega M}\left\{ \sum_{m=0}^{M-1} e^{-j\omega r_m T} e^{-jkm\frac{2\pi}{M}} - e^{-j\omega T} \sum_{m=0}^{M-1} e^{-j\omega r_{m+1}T} e^{-jkm\frac{2\pi}{M}} \right\} \quad (A1.32)$$

At $\omega = \omega_o + k(2\pi)/MT$, (A1.32) becomes

$$A_{k,IU-ON(SH)}(\omega_o + k\frac{2\pi}{MT}) \approx \frac{1}{j(\omega_o + k\frac{2\pi}{MT})M}\left\{\sum_{m=0}^{M-1}e^{-jkm\frac{2\pi}{M}} - j(\omega_o T + k\frac{2\pi}{M})\sum_{m=0}^{M-1}r_m e^{-jkm\frac{2\pi}{M}}\right.$$

$$\left. -e^{-j\omega_o T}\left[\sum_{m=0}^{M-1}e^{-jk(m+1)\frac{2\pi}{M}} - j(\omega_o T + k\frac{2\pi}{M})\sum_{m=0}^{M-1}r_{m+1}e^{-jk(m+1)\frac{2\pi}{M}}\right]\right\}$$

$$(A1.33)$$

as r_m & $e^{-jkm(2\pi)/M}$ is periodic with period $m = M$, so

$$\sum_{m=0}^{M-1}e^{-jk(m+1)\frac{2\pi}{M}} = \sum_{m=1}^{M}e^{-jkm\frac{2\pi}{M}} = \sum_{m=0}^{M-1}e^{-jkm\frac{2\pi}{M}}$$

and $\quad \sum_{m=0}^{M-1}r_{m+1}e^{-jk(m+1)\frac{2\pi}{M}} = \sum_{m=1}^{M}r_m e^{-jkm\frac{2\pi}{M}} = \sum_{m=0}^{M-1}r_m e^{-jkm\frac{2\pi}{M}}$

where (A1.33) can be simplified to

$$A_{k,IU-ON(SH)}(\omega_o + k\frac{2\pi}{MT}) \approx -\frac{2jT}{M}\sin\left(\frac{\omega_o T}{2}\right)e^{-j\omega_o T/2}\sum_{m=0}^{M-1}r_m e^{-jkm\frac{2\pi}{M}} \qquad \text{for} \quad k \neq 0$$

$$(A1.34)$$

Again $A_{k,IU-ON(SH)}$ for $k=0$ is calculated separately. Substitute $\omega = \omega_o$ and $k=0$ into (A1.33) yields:

$$A_{0,IU-ON(SH)}(\omega_o) \approx \frac{2}{\omega_o}\sin\left(\frac{\omega_o T}{2}\right)e^{-j\omega_o T/2} \qquad (A1.35)$$

Similarly, it is easy to find the $A_{k,IU-ON(SH)}$ produced by the input component $\pi\delta(\Omega + \Omega_o)$ at frequency $\omega = -\omega_o + 2\pi(M-k)/MT$, which are following:

$$A_{k,IU-ON(SH)}(-\omega_o + (M-k)\frac{2\pi}{MT}) \approx \frac{2jT}{M}\sin\left(\frac{\omega_o T}{2}\right)e^{j\omega_o T/2}\sum_{m=0}^{M-1}r_m e^{jkm\frac{2\pi}{M}} \quad \text{for} \ k \neq 0 \ (A1.36)$$

$$A_{0,IU-ON(SH)}(\omega_s - \omega_0) \approx \frac{-2}{\omega_s - \omega_o} \sin\left(\frac{\omega_0 T}{2}\right) e^{j\omega_0 T/2} \qquad (A1.37)$$

The summary of the equations derived is as follows:
For IN-OU(IS),

$$A_{k,IN-OU(IS)}(\omega) = \begin{cases} \dfrac{j\omega_o T}{M} \displaystyle\sum_{m=0}^{M-1} r_m e^{-jkm\frac{2\pi}{M}}, & for\ \omega = \omega_o + k\dfrac{2\pi}{MT}\ and\ k \neq 0; \\[4mm] 1, & for\ \omega = \omega_o; \\[2mm] -\dfrac{j\omega_o T}{M} \displaystyle\sum_{m=0}^{M-1} r_m e^{jkm\frac{2\pi}{M}}, & for\ \omega = -\omega_o + (M-k)\dfrac{2\pi}{MT}\ and\ k \neq 0; \\[4mm] 1, & for\ \omega = \omega_s - \omega_o; \end{cases}$$

$$(A1.38)$$

and for IU-ON(SH),

$$A_{k,IU-ON(SH)}(\omega) = \begin{cases} -\dfrac{2jT}{M} \sin\left(\dfrac{\omega_o T}{2}\right) e^{-j\omega_o T/2} \displaystyle\sum_{m=0}^{M-1} r_m e^{-jkm\frac{2\pi}{M}}, & \\[4mm] & for\ \omega = \omega_o + k\dfrac{2\pi}{MT}\ and\ k \neq 0; \\[2mm] \dfrac{2}{\omega_o} \sin\left(\dfrac{\omega_o T}{2}\right) e^{-j\omega_o T/2}, & for\ \omega = \omega_o; \\[4mm] \dfrac{2jT}{M} \sin\left(\dfrac{\omega_o T}{2}\right) e^{j\omega_o T/2} \displaystyle\sum_{m=0}^{M-1} r_m e^{jkm\frac{2\pi}{M}}, & \\[4mm] & for\ \omega = -\omega_o + (M-k)\dfrac{2\pi}{MT}\ and\ k \neq 0; \\[2mm] \dfrac{-2}{\omega_s - \omega_o} \sin\left(\dfrac{\omega_o T}{2}\right) e^{j\omega_o T/2}, & for\ \omega = \omega_s - \omega_o; \end{cases}$$

$$(A1.39)$$

Magnitude Relationship:
The magnitude of $A_{k,IN-OU(IS)}(\omega)$ can be expressed as:

$$\left| A_{k,IN-OU(IS)}(\omega) \right| = \begin{cases} \dfrac{\omega_o T}{M} \left| \displaystyle\sum_{m=0}^{M-1} r_m e^{-jkm\frac{2\pi}{M}} \right|, & for\ \omega = \omega_o + k\dfrac{2\pi}{MT}\ and\ k \neq 0; \\[4mm] 1, & for\ \omega = \omega_o; \\[2mm] \dfrac{\omega_o T}{M} \left| \displaystyle\sum_{m=0}^{M-1} r_m e^{jkm\frac{2\pi}{M}} \right|, & for\ \omega = -\omega_o + (M-k)\dfrac{2\pi}{MT}\ and\ k \neq 0; \\[4mm] 1, & for\ \omega = \omega_s - \omega_o; \end{cases}$$

$$(A1.40)$$

The normalized magnitude with respect to the signal component at $\omega = \omega_0$ and this component is equal to 1, so (A1.40) also represents normalized $\left| A_{k,IN-OU(IS)}(\omega) \right|$.

The magnitude of $A_{k,IU-ON(SH)}(\omega)$ can be expressed as:

$$\left| A_{k,IU-ON(SH)}(\omega) \right| = \begin{cases} \left| \dfrac{2T}{M} \sin\left(\dfrac{\omega_0 T}{2} \right) \displaystyle\sum_{m=0}^{M-1} r_m e^{-jkm\frac{2\pi}{M}} \right|, & \text{for } \omega = \omega_0 + k\dfrac{2\pi}{MT} \text{ and } k \neq 0; \\[2ex] \dfrac{2}{\omega_0} \sin\left(\dfrac{\omega_0 T}{2} \right), & \text{for } \omega = \omega_0; \\[2ex] \left| \dfrac{2T}{M} \sin\left(\dfrac{\omega_0 T}{2} \right) \displaystyle\sum_{m=0}^{M-1} r_m e^{jkm\frac{2\pi}{M}} \right|, & \text{for } \omega = -\omega_0 + (M-k)\dfrac{2\pi}{MT} \text{ and } k \neq 0; \\[2ex] \dfrac{2}{\omega_s - \omega_0} \sin\left(\dfrac{\omega_0 T}{2} \right), & \text{for } \omega = \omega_s - \omega_0; \end{cases}$$

(A1. 41)

and the normalized $\left| A_{k,IU-ON(SH)}(\omega) \right|$ with respect to the signal component at $\omega = \omega_0$ is

$$\left| A_{k,IU-ON(SH),normalized}(\omega) \right| = \begin{cases} \left| \dfrac{\omega_0 T}{M} \displaystyle\sum_{m=0}^{M-1} r_m e^{-jkm\frac{2\pi}{M}} \right|, & \text{for } \omega = \omega_0 + k\dfrac{2\pi}{MT} \text{ and } k \neq 0; \\[2ex] 1, & \text{for } \omega = \omega_0; \\[2ex] \left| \dfrac{\omega_0 T}{M} \displaystyle\sum_{m=0}^{M-1} r_m e^{jkm\frac{2\pi}{M}} \right|, & \text{for } \omega = -\omega_0 + (M-k)\dfrac{2\pi}{MT} \text{ and } k \neq 0; \\[2ex] \dfrac{\omega_0}{\omega_s - \omega_0}, & \text{for } \omega = \omega_s - \omega_0; \end{cases}$$

(A1.42)

Comparing (A1.40) & (A1.42), it is obvious to deduct that the normalized distortion components in the two cases are the same for arbitrary parallel path number M, so that all the formulas previously derived [3.5, 3.6, 3.7, 3.8, 3.9, 3.10] for evaluating the SINAD for IN-OU(IS), e.g. the formula (3.23), also can be applied for IU-ON(SH). This spectrum similarity is valid also for $M \to \infty$ which is the case of pure random sampling jitter, thus this also allows us to make use of formulas (3.24) and (3.25) to estimate the noise floor for the case of pure random sampling jitter.

Simulation results are shown in Figure A1-8 to verify our theoretical derivation. Figure A1-8(a) shows mean SINAD for the output signals of IU-ON(SH) systems, while Figure A1-8(b) presents the relative difference of mean SINAD between the output of the two cases. The results show that for reasonably large SINAD (>30 dB for $\sigma_{rm} f_0 / f_s < 0.5\%$), this relative difference is well below 0.1%, which match our derivation very well. Also, the relative difference increases when either jitter standard variation or signal frequency increases.

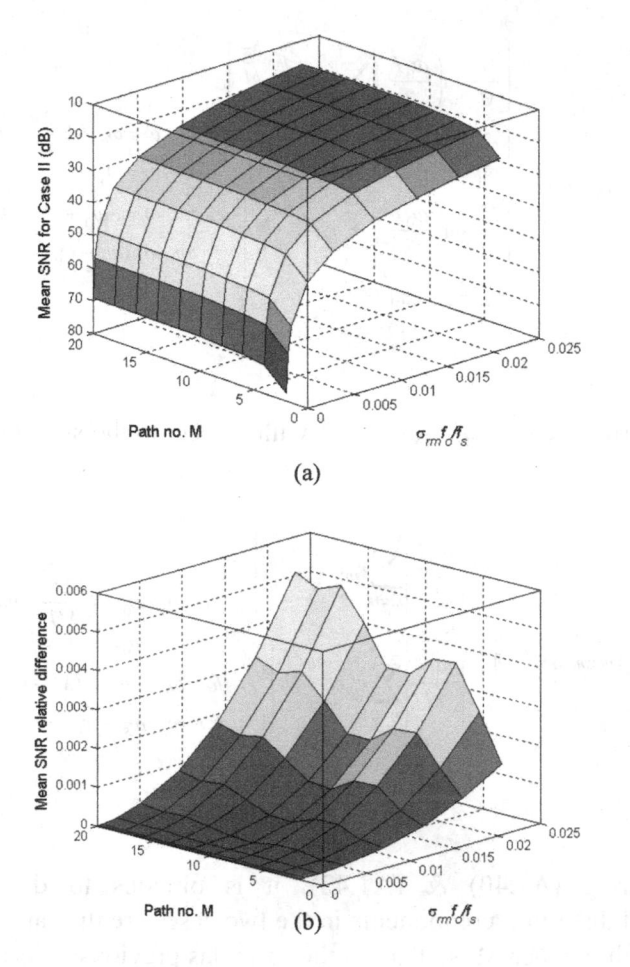

Figure A1-8. (a) Mean SINAD for IU-ON(SH) and (b) Relative difference of Mean SINAD between IN-OU(IS) & IU-ON(SH) versus signal frequency, standard derivation of skew-timing ratio r_m and the path number M

Phase Relationship:

In additional to magnitude relationship, certain phase relationship can also be derived. The phase of $A_{k,IN-OU(IS)}(\omega)$ can be expressed as:

$$\angle A_{k,IN-OU(IS)}(\omega) = \begin{cases} \dfrac{\pi}{2} + \angle\left(\displaystyle\sum_{m=0}^{M-1} r_m e^{-jkm\frac{2\pi}{M}}\right), & \text{for } \omega = \omega_o + k\dfrac{2\pi}{MT} \text{ and } k \neq 0; \\[4mm] 0, & \text{for } \omega = \omega_o; \\[2mm] -\dfrac{\pi}{2} + \angle\left(\displaystyle\sum_{m=0}^{M-1} r_m e^{jkm\frac{2\pi}{M}}\right), & \text{for } \omega = -\omega_o + (M-k)\dfrac{2\pi}{MT} \text{ and } k \neq 0; \\[4mm] 0, & \text{for } \omega = \omega_s - \omega_o; \end{cases}$$

$$\text{(A1.43)}$$

The phase of $A_{k,IU-ON(SH)}(\omega)$ can be expressed as:

$$\angle A_{k,IU-ON(SH)}(\omega) = \begin{cases} -\dfrac{\pi}{2} - \dfrac{\omega_o T}{2} + \angle\left(\displaystyle\sum_{m-0}^{M-1} r_m e^{-jkm\frac{2\pi}{M}}\right), & \\[2mm] & \text{for } \omega = \omega_o + k\dfrac{2\pi}{MT} \text{ and } k \neq 0; \\[4mm] -\dfrac{\omega_o T}{2}, & \text{for } \omega = \omega_o; \\[4mm] \dfrac{\pi}{2} + \dfrac{\omega_o T}{2} + \angle\left(\displaystyle\sum_{m=0}^{M-1} r_m e^{jkm\frac{2\pi}{M}}\right), & \text{for } \omega = -\omega_o + (M-k)\dfrac{2\pi}{MT} \text{ and } k \neq 0; \\[4mm] & \text{for } \omega = \omega_s - \omega_o; \\[2mm] \pi + \dfrac{\omega_o T}{2}, & \end{cases}$$

$$\text{(A1.44)}$$

Finally we can find the phase difference between $A_{k,IN-OU(IS)}(\omega)$ & $A_{k,IU-ON(SH)}(\omega)$ by subtracting (A1.43) from (A1.44):

$$\angle A_{k,IU-ON(SH)}(\omega) - \angle A_{k,IN-OU(IS)}(\omega) = \begin{cases} -\pi - \dfrac{\omega_o T}{2}, & \\[2mm] & \text{for } \omega = \omega_o + k\dfrac{2\pi}{MT} \text{ and } k \neq 0; \\[4mm] -\dfrac{\omega_o T}{2}, & \text{for } \omega = \omega_o; \\[4mm] \pi + \dfrac{\omega_o T}{2}, & \text{for } \omega = -\omega_o + (M-k)\dfrac{2\pi}{MT} \text{ and } k \neq 0; \\[4mm] \pi + \dfrac{\omega_o T}{2}, & \text{for } \omega = \omega_s - \omega_o; \end{cases}$$

$$\text{(A1.45)}$$

Appendix 2

NOISE ANALYSIS FOR SC ADB DELAY LINE AND POLYPHASE SUBFILTERS

1. OUTPUT NOISE OF ADB DELAY LINE

For the ADB in the delay line (Figure 3-10 (a) from chapter 3), mainly the thermal noise contributes for phase A, i.e., $\overline{v_{n1}^2} = kT/C_s$, produced by on-resistance of the switch of phase A and stored in sampling capacitor C_s. Thus, the output noise power contribution due to this phase is equal to

$$\overline{v_{no,ADB,A}^2} = \frac{kT}{C_s} \tag{A2.1}$$

At phase 2 (Figure 3-10 (b)), there are two additional noise contributions: $\overline{v_{ns}^2}$, the noise produced by the on-resistance of the switch in phase B, and $\overline{v_{nOA}^2}$, the noise generated by the transistor inside the opamp. These noises will be shaped by the closed-loop opamp bandwidth, assuming opamp as a single-pole system, the transfer function from the noise source $\overline{v_{ns}^2}$ to the output is

$$H_s(s) = \frac{1}{1 + \dfrac{s}{\omega_{ugb\_d}\beta_d}} \tag{A2.2}$$

where the feedback factor and unity-gain bandwidth of opamp are given by

$$\beta_d = \frac{C_s}{C_s + C_{PI\_d}}$$ (A2.3)

and

$$\omega_{ugb\_d} = \frac{g_{m1\_d}}{C_{L,tot\_d}} = \frac{g_{m1\_d}\left(C_s + C_{PI\_d}\right)}{C_{L\_d}\left(C_s + C_{PI\_d}\right) + C_s C_{PI\_d}}$$ (A2.4)

Considering the equivalent noise bandwidth for 1st-order as

$$BW_N = \frac{1}{4}\beta_d \omega_{ugb\_d}$$ (A2.5)

the output noise power due to noise source $\overline{v_{ns}^2}$ is

$$\overline{v_{no,ADB,B,s}^2} = S_s(f)\cdot|H_s(0)|^2 \cdot BW_N = kTR_s\omega_{ugb\_d}\cdot\frac{C_s}{C_s + C_{PI\_d}}$$ (A2.6)

The transfer function from opamp noise source $\overline{v_{nOA}^2}$ to output is

$$H_{OA1}(s) = \frac{1}{\beta_d}\cdot\frac{1}{1 + \dfrac{s}{\omega_{ugb\_d}\beta_d}}$$ (A2.7)

and the noise power spectral density for the single-stage opamp is

$$S_{OA1}(f) = \frac{8}{3}\frac{kT}{g_{m1\_d}}\gamma_d$$ (A2.8)

where γ_d represents the excess noise factor for opamp in ADB. Thus, the output noise power due to noise source $\overline{v_{nOA}^2}$ is given by

$$\overline{v_{no,ADB,B,OA1}^2} = S_{OA1}(f)\cdot|H_{OA1}(0)|^2 \cdot BW_N = \frac{2}{3}\frac{kT}{g_{m1\_d}}\gamma_d\omega_{ugb\_d}\frac{C_s + C_{PI\_d}}{C_s}$$ (A2.9)

The total noise power for the ADB single positive/negative output can then be represented as follows:

$$\overline{v_{no,ADB,1/2}^2} = \overline{v_{no,ADB,A}^2} + \overline{v_{no,ADB,B,s}^2} + \overline{v_{no,ADB,B,OA1}^2}$$

$$= \frac{kT}{C_s} + \left[\left(\frac{kTR_sC_s}{C_s + C_{PI\_d}}\right) + \frac{2}{3}\frac{kT}{g_{m1\_d}}\gamma_d\left(\frac{C_s + C_{PI\_d}}{C_s}\right)\right]\omega_{ugb\_d} \qquad (A2.10)$$

Considering that the same opamp is used for all ADB's, the total output noise for the delay line with *i*-number of ADBs in phase B is derived as

$$\overline{v_{no,ADBi,1/2}^2} = i \cdot \overline{v_{no,ADB,1/2}^2} \qquad (A2.11)$$

i.e. the total output noise power of ADB delay line is proportional to the numbers of the ADB.

2. OUTPUT NOISE OF POLYPHASE SUBFILTERS

2.1 Using TSI Input Coefficient SC Branches

For the *L*-path polyphase subfilter (Figure 3-11), two kind of noise sources contribute at phase A: $v_{ni\_A}^2 = kT/C_i$, the noise produced by the on-resistance of ith input path, and $v_{nF\_A}^2 = kT/C_F$, the noise produced by the on-resistance at the reset path of C_F. The noise power from input will be scaled according to the capacitor ratio of C_i to C_F. The noise charge stored in capacitor C_i in phase A can be calculated as

$$Q_{ni}^2 = C_i^2\left(\frac{kT}{C_i}\right) = kTC_i \qquad (A2.12)$$

The charge from all the input C_i's will be transferred to C_F in next phase. Thus, its total equivalent noise that will contribute to the output in phase B is

$$\overline{v^2_{no,PFm,A}} = \frac{1}{C_F^2}\sum_{i=1}^{n} Q_{ni}^2 + \frac{kT}{C_F} = \frac{kT}{C_F^2}\sum_{i=1}^{n} C_i + \frac{kT}{C_F} \tag{A2.13}$$

At phase B, the feedback factor is

$$\beta_{PF} = \frac{C_F}{C_F + \displaystyle\sum_{i=1}^{n} C_i + C_{PI\_PF}} \tag{A2.14}$$

and the unity-gain bandwidth for this stage is

$$\omega_{ugb\_PF} = \frac{g_{m1\_PF}}{C_{L,tot\_PF}}$$

$$= \frac{g_{m1\_PF}\left(C_F + \displaystyle\sum_{i=1}^{n} C_i + C_{PI\_PF}\right)}{C_{L\_PF}\left(C_F + \displaystyle\sum_{i=1}^{n} C_i + C_{PI\_PF}\right) + C_F\left(\displaystyle\sum_{i=1}^{n} C_i + C_{PI\_PF}\right)} \tag{A2.15}$$

Three kind of additional noise sources contribute to output in this phase (Figure 3-11 (b)): $\overline{v^2_{ni}}$, $\overline{v^2_{nF}}$, and $\overline{v^2_{nOA}}$. The transfer function from noise source $\overline{v^2_{ni}}$ to the output of this stage is

$$H_i(s) = -\frac{C_i}{C_F} \cdot \frac{1}{1 + \dfrac{s}{\omega_{ugb\_PF}\beta_{PF}}} \tag{A2.16}$$

Then the output noise power due to the noise source $\overline{v^2_{ni}}$ is

$$\overline{v^2_{no,PFm,B,i}} = S_i(f)\cdot\left|H_i(0)\right|^2 \cdot BW_N = \frac{kTR_i C_i^2 \omega_{ugb\_PF}}{C_F\left(C_F + \displaystyle\sum_{i=1}^{n} C_i + C_{PI\_PF}\right)} \tag{A2.17}$$

The case for the noise source $\overline{v^2_{nF}}$ for this stage is similar to the noise source $\overline{v^2_{ns}}$ in the ADB's calculation, and the output noise power due to this noise source is

$$\overline{v_{no,PFm,B,F}^2} = kTR_F\omega_{ugb\_PF} \cdot \frac{C_F}{C_F + \sum_{i=1}^{n}C_i + C_{PI\_PF}} \qquad (A2.18)$$

The transfer function from the noise source $\overline{v_{nOA}^2}$ to the output and the noise power spectral density $S_{OA2}(f)$ is the same as (A2.7) and (A2.8), and the output noise power due to the noise source $\overline{v_{nOA}^2}$ is given by

$$\overline{v_{no,PFm,B,OA2}^2} = S_{OA2}(f)\cdot|H_{OA2}(0)|^2\cdot BW_N = \frac{2}{3}\frac{kT}{g_{m1\_PF}}\gamma_{PF}\,\omega_{ugb\_PF}\frac{C_F + \sum_{i=1}^{n}C_i + C_{PI\_PF}}{C_F} \qquad (A2.19)$$

Therefore, the total noise power for the polyphase subfilter positive/negative output can then be found as:

$$\overline{v_{no,PFm,1/2}^2} = \overline{v_{no,PFm,A}^2} + \sum_{i=1}^{n}\overline{v_{no,PFm,B,i}^2} + \overline{v_{no,PFm,B,F}^2} + \overline{v_{no,PFm,B,OA2}^2}$$

$$= \frac{kT}{C_F^2}\sum_{i=1}^{n}C_i + \frac{kT}{C_F} + \left(\frac{kT\sum_{i=1}^{n}R_iC_i^2}{C_F(C_F + \sum_{i=1}^{n}C_i + C_{PI\_PF})}\right)\omega_{ugb\_PF}$$

$$+ \left(\frac{kTR_FC_F}{C_F + \sum_{i=1}^{n}C_i + C_{PI\_PF}}\right)\omega_{ugb\_PF}$$

$$+ \frac{2}{3}\frac{kT}{g_{m1\_PF}}\left(1 + \frac{g_{m7\_PF}}{g_{m1\_PF}}\right)\left(\frac{C_F + \sum_{i=1}^{n}C_i + C_{PI\_PF}}{C_F}\right)\omega_{ugb\_PF} \qquad (A2.20)$$

2.2 Using OFR Input Coefficient SC Branches

The derivation shown above is applied to TSI input branches: sample the input signal in capacitor C_i at phase A and transfer the charge to output at phase B. Nevertheless, for flexibly allocating the delays needed for different tap weights, some coefficients are implemented also by OFR SC branches: C_i directly couple the input to output in phase B, while being reset at phase A. However, the noise contributions will be the same as that using TSI branches: at phase A, the noise will be produced by the thermal on-resistance of ith input path and the on-resistance at the reset path of C_F; while at phase B, the total output noise contribution is derived from the charge-redistribution of the entire stored noise at phase A, the switch thermal on-resistances in the whole charge transferring path as well as the opamp input-referred noise, as presented in (A2.20).

Appendix 3

GAIN, PHASE AND OFFSET ERRORS FOR GOC MF SC DELAY CIRCUIT I AND J

1. GOC MF SC DELAY CIRCUIT I[1]

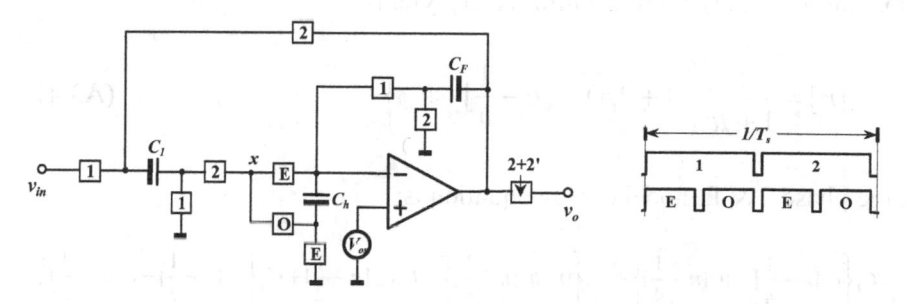

Figure A3 -1. EC/P-CDS GOC MF SC delay circuit (i)

Assuming $C_1/C_F = k_1$, $C_h/C_F = k_h$, $C_h/C_1 = k_h'$, $\mu = 1/A$, then at phases 2 & O, the nodal charge equation yields:

$$C_1\{v_o[n] - v_x[n]\} + C_h\{v_-[n] - v_x[n]\} = C_1\left\{v_o[n-\tfrac{1}{4}] - v_-[n-\tfrac{1}{4}]\right\} + C_h\left\{v_-[n-\tfrac{1}{4}] - 0\right\}$$

$$(A3.1)$$

---

[1] Since the derivation of the gain, phase and offset of the SC delay circuits is lengthy, for simplicity we present here only for two circuits, and all the others can be similarly derived according to the presented methodology.

where

$$v_-[n] = -\mu v_o[n] + V_{os} \tag{A3.2}$$

$v_x[n]$ is found as follows: at phase 2 & E, the charge stored in C_h is $C_h v_-[n-1/4]$, while at phase 2 & O,

$$C_h\{v_x[n] - v_-[n]\} = C_h v_-[n - \frac{1}{4}]$$

and $v_x[n] + \dfrac{v_o[n]}{A} - V_{os} = \dfrac{v_o[n - \frac{1}{4}]}{A} - V_{os}$

$$\therefore v_x[n] = -\mu\left\{v_o[n] - v_o[n - \frac{1}{4}]\right\} \tag{A3.3}$$

Substituting (A3.2) & (A3.3) into (A3.1), yields

$$v_o[n] = \frac{1}{1+\mu}\left\{(1+2\mu)v_o[n - \frac{1}{4}] - V_{os}\right\} \tag{A3.4}$$

At the phase 2 & E, nodal charge equation is:

$$C_1\left\{v_o[n-\frac{1}{4}] - v_-[n-\frac{1}{4}]\right\} + C_h\left\{0 - v_-[n-\frac{1}{4}]\right\} = C_1 v_{in}[n-\frac{1}{2}] + C_h\left\{v_x[n-\frac{1}{2}] - v_-[n-\frac{1}{2}]\right\} \tag{A3.5}$$

Substituting (A3.2) & (A3.3) into (A3.5), yields

$$v_o[n-\frac{1}{4}] = \frac{1}{1+\mu(1+k'_h)}\left\{v_{in}[n-\frac{1}{2}] + V_{os} + k'_h\mu v_o[n-\frac{3}{4}]\right\} \tag{A3.6}$$

Substitute (A3.6) into (A3.4), $v_o[n]$ is obtained by:

$$v_o[n] = \frac{1+2\mu}{(1+\mu)[1+\mu(1+k'_h)]}v_{in}[n-\frac{1}{2}] + \frac{k'_h\mu(1+2\mu)}{(1+\mu)[1+\mu(1+k'_h)]}v_o[n-\frac{3}{4}]$$

$$+\frac{(1+2\mu)-[1+\mu(1+k'_h)]}{(1+\mu)[1+\mu(1+k'_h)]}V_{os} \tag{A3.7}$$

Similarly, write the nodal charge equation at the phase 1 & O and phase 1 & E, after the simplification, we obtain

$$v_o[n-\frac{3}{4}]=\frac{1}{1+\mu(1+k_h)}v_o[n-1]+\frac{k_h\mu}{1+\mu(1+k_h)}v_o[n-\frac{5}{4}]+\frac{1}{1+\mu(1+k_h)}V_{os} \tag{A3.8}$$

Substitute (A3.8) into (A3.7) yields

$$v_o[n]=\frac{1+2\mu}{mn}v_{in}[n-\frac{1}{2}]+\frac{k'_h\mu(1+2\mu)}{mnp}v_o[n-1]+\frac{k'_hk_h\mu^2(1+2\mu)}{mnp}v_o[n-\frac{5}{4}]$$

$$+\frac{k'_h\mu(1+2\mu)+[(1+2\mu)-n]p}{mnp}V_{os} \tag{A3.9}$$

where $m=(1+\mu)$, $n=1+\mu(1+k'_h)$, $p=1+\mu(1+k_h)$
By reapplying (A3.6) and (A3.8), we finally obtain

$$v_o[n-\frac{5}{4}]=\frac{v_{in}[n-\frac{3}{2}]}{m}+\frac{k'_h\mu}{mp}v_o[n-2]+\frac{k'_hk_h\mu^2}{mp}v_o[n-\frac{9}{4}]+\frac{k'_h\mu}{mp}V_{os}+\frac{V_{os}}{m} \tag{A3.10}$$

Finally, by substituting (A3.10) into (A3.9) the output can be expressed as

$$v_o[n]=\frac{1+2\mu}{mn}v_{in}[n-\frac{1}{2}]+\frac{k'_h\mu(1+2\mu)}{mnp}v_o[n-1]+\frac{k'_hk_h\mu^2(1+2\mu)}{m^2np}v_{in}[n-\frac{3}{2}]$$

$$+\frac{k'^2_hk_h\mu^3(1+2\mu)}{m^2np^2}v_o[n-2]+\frac{k'^2_hk^2_h\mu^4(1+2\mu)}{m^2np^2}v_o[n-\frac{9}{4}]$$

$$+\frac{k'_hk_h\mu^2(1+2\mu)(k'_h\mu+p)}{m^2np^2}V_{os}+\frac{k'_h\mu(1+2\mu)+[(1+2\mu)-n]p}{mnp}V_{os} \tag{A3.11}$$

Therefore, when $v_{in}[n] = 0$, the offset suppression factor is given by

$$\gamma = \frac{k'_h k_h \mu^2 (1+2\mu)(k'_h \mu + p)}{m^2 n p^2} + \frac{k'_h \mu(1+2\mu) + [(1+2\mu) - n]p}{mnp}$$

$$\approx \frac{k'_h \mu(1+2\mu) + \{(1+2\mu) - [1+\mu(1+k'_h)]\}[1+\mu(1+k_h)]}{(1+\mu)[1+\mu(1+k'_h)][1+\mu(1+k_h)]} \quad (A3.12)$$

Let the offset voltage be zero, and the transfer function can be derived as

$$H(z) = \frac{(1+2\mu)z^{-1/2} + \dfrac{k'_h k_h \mu^2 (1+2\mu)}{mp} z^{-3/2}}{mn - \dfrac{k'_h \mu(1+2\mu)}{p} z^{-1} - \dfrac{k'^2_h k_h \mu^3 (1+2\mu)}{mp^2} z^{-2}} \quad (A3.13)$$

For an ideal op amp, $\mu = 0$, it gives the ideal delay circuit transfer function $H_D(z) = z^{-1/2}$. And the Fourier transform of (A3.13) can be approximately expressed as

$$H(e^{j\omega}) = \frac{(1+2\mu)e^{-j\omega/2}}{mn - \dfrac{k'_h \mu(1+2\mu)}{p} e^{-j\omega}} = \frac{e^{-j\omega/2}}{1 - m(e^{j\omega}) - j\theta(e^{j\omega})} \quad (A3.14)$$

where $m(e^{j\omega})$, $\theta(e^{j\omega})$ represents the gain and phase error respectively [4.27], which are expressed by

$$m(e^{j\omega}) = -\frac{\mu[k'_h + \mu(1+k'_h)]}{1+2\mu} + \frac{k'_h \mu}{1+\mu(1+k_h)}\cos\omega \quad (A3.15)$$

$$\theta(e^{j\omega}) = -\frac{k'_h \mu}{1+\mu(1+k_h)}\sin\omega \quad (A3.16)$$

2. GOC MF SC DELAY CIRCUIT J

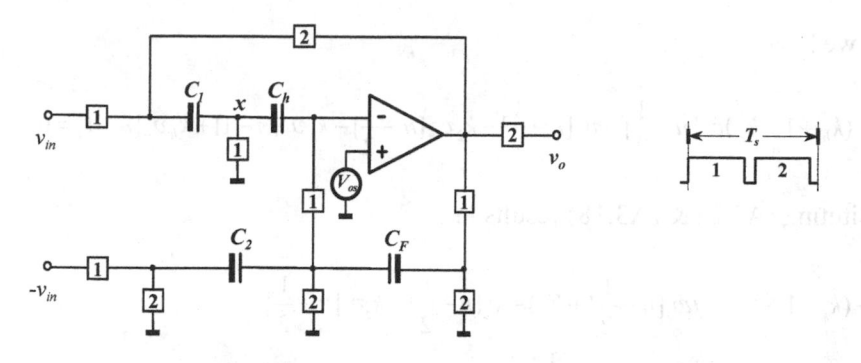

Figure A3-2. Differential-input, EC/P-CDS GOC MF SC delay circuit (j)

Assuming $C_1/C_F = k_1$, $C_h/C_F = k_h$, $C_2/C_F = k_2$, $\mu = 1/A$, then at the phase 2, the nodal charge equation yields:

$$C_1\{v_o[n] - v_x[n]\} + C_h\{v_-[n] - v_x[n]\} = C_1\left\{v_{in}[n - \tfrac{1}{2}] - 0\right\} + C_h\left\{v_-[n - \tfrac{1}{2}] - 0\right\}$$

$$\text{(A3.17)}$$

where $v_-[n]$ is also given by (A3.2). $v_x[n]$ can be found also similarly as (A3.3), i.e.

$$v_x[n] = -\mu v_o[n] + \mu v_o[n - \tfrac{1}{2}] \qquad\qquad \text{(A3.18)}$$

and substitute (A3.2), (A3.18) into (A3.17), after simplification yields

$$(1 + \mu)v_o[n] = \mu v_o[n - \tfrac{1}{2}] + v_{in}[n - \tfrac{1}{2}] \qquad\qquad \text{(A3.19)}$$

At the phase 1, the nodal charge equations at the inverting node of op amp is:

$$C_h\left\{0 - v_-[n - \tfrac{1}{2}]\right\} - C_h\{v_x[n - 1] - v_-[n - 1]\} + C_2\left\{-v_{in}[n - \tfrac{1}{2}] - v_-[n - \tfrac{1}{2}]\right\} - C_2\{0\}$$

$$+ C_F \left\{ v_o[n - \frac{1}{2}] - v_-[n - \frac{1}{2}] \right\} - C_F\{0\} = 0$$

Then we have

$$- (k_h + 1 + k_2) v_-[n - \frac{1}{2}] + v_o[n - \frac{1}{2}] - k_2 v_{in}[n - \frac{1}{2}] - k_h v_x[n-1] + k_h v_-[n-1] = 0$$

Substituting (A3.2) & (A3.18) results in

$$- (k_h + 1 + k_2)(-\mu v_o[n - \frac{1}{2}] + V_{os}) + v_o[n - \frac{1}{2}] - k_2 v_{in}[n - \frac{1}{2}]$$

$$- k_h(-\mu v_o[n-1] + \mu v_o[n - \frac{3}{2}]) + k_h(-\mu v_o[n-1] + V_{os}) = 0$$

Solve for $v_o[n - 1/2]$, yields

$$v_o[n - \frac{1}{2}] = \frac{1}{1 + \mu(k_h + k_2 + 1)} \left\{ k_2 v_{in}[n - \frac{1}{2}] + \mu k_h v_o[n - \frac{3}{2}] + (k_2 + 1)V_{os} \right\}$$

$$(A3.20)$$

By applying (A3.20) recursively, we obtain

$$v_o[n - \frac{1}{2}] = \frac{k_2}{1 + \mu(k_h + k_2 + 1)} v_{in}[n - \frac{1}{2}] + \frac{(k_2 + 1)[1 + \mu(2k_h + k_2 + 1)]}{[1 + \mu(k_h + k_2 + 1)]^2} V_{os}$$

$$+ \frac{\mu k_h k_2}{[1 + \mu(k_h + k_2 + 1)]^2} v_{in}[n - \frac{3}{2}] + \frac{\mu^2 k_h^2}{[1 + \mu(k_h + k_2 + 1)]^2} v_o[n - \frac{5}{2}]$$

$$\approx \frac{k_2}{1 + \mu(k_h + k_2 + 1)} v_{in}[n - \frac{1}{2}] + \frac{\mu k_h k_2}{[1 + \mu(k_h + k_2 + 1)]^2} v_{in}[n - \frac{3}{2}]$$

$$+ \frac{(k_2 + 1)[1 + \mu(2k_h + k_2 + 1)]}{[1 + \mu(k_h + k_2 + 1)]^2} V_{os} \qquad (A3.21)$$

By substituting (A3.21) into (A3.19) and solve for $v_o[n]$ it yields:

$$v_o[n] = \frac{\mu(k_h + 2k_2 + 1) + 1}{[1 + \mu(k_h + k_2 + 1)](1 + \mu)} v_{in}[n - \frac{1}{2}] + \frac{\mu^2 k_h k_2}{[1 + \mu(k_h + k_2 + 1)]^2 (1 + \mu)} v_{in}[n - \frac{3}{2}]$$

$$+\frac{\mu(k_2+1)[1+\mu(2k_h+k_2+1)]}{[1+\mu(k_h+k_2+1)]^2(1+\mu)}V_{os} \qquad\text{(A3.22)}$$

So when $v_{in}[n]=0$, the offset suppression factor is given by

$$\gamma=\frac{\mu(k_2+1)[1+\mu(2k_h+k_2+1)]}{[1+\mu(k_h+k_2+1)]^2(1+\mu)} \qquad\text{(A3.23)}$$

Let $V_{os}=0$, then the transfer function for the delay circuit will be

$$H(z)=\frac{z^{-1/2}}{[1+\mu(k_h+k_2+1)](1+\mu)}\left\{1+\mu(k_h+2k_2+1)+\frac{\mu^2k_hk_2}{1+\mu(k_h+k_2+1)}z^{-1}\right\}$$

$$\text{(A3.24)}$$

For an ideal op amp, $\mu=0$, it gives the ideal delay circuit transfer function $H_D(z)=z^{-1/2}$, and the Fourier transform of (A3.25) is expressed as

$$H(e^{j\omega})=\frac{e^{-j\omega/2}}{[1+\mu(k_h+k_2+1)](1+\mu)}\left\{1+\mu(k_h+2k_2+1)+\frac{\mu^2k_hk_2}{1+\mu(k_h+k_2+1)}e^{-j\omega}\right\}$$

$$\text{(A3.25)}$$

By ignoring the terms with factor of μ^2, the gain and phase errors can be approximately obtained by

$$m(e^{j\omega})=\frac{\mu^2k_hk_2\cos\omega}{[1+\mu(k_h+k_2+1)][1+\mu(k_h+2k_2+1)]}-\frac{\mu[1-k_2+\mu(k_h+k_2+1)]}{1+\mu(k_h+2k_2+1)}$$

$$\text{(A3.26)}$$

$$\theta(e^{j\omega})=\frac{-\mu^2k_hk_2\sin\omega}{[1+\mu(k_h+k_2+1)][1+\mu(k_h+2k_2+1)]}$$